TEACHERS HELPING TEAC

Mathematics, Science, & Technology Connections

Project Co-ordinator: Bob Corney
Project Manager: John Zelisko
Writing Team: Peel Board of Education Teachers

TRIFOLIUM BOOKS INC.

Trifolium Books Inc.
250 Merton Street, Suite 203
Toronto, Ontario, Canada M4S 1B1

Canadian Cataloguing in Publication Data
Corney, Bob, 1934-
Mathematics, science, & technology connections

(Teachers helping teachers)
Includes bibliographical references.
ISBN 1-895579-37-6

1. Science - Study and teaching (Elementary).
2. Science - Study and teaching (Secondary).
3. Mathematics - Study and teaching (Elementary). 4. Mathematics - Study and teaching (Secondary). 5. Technology - Study and teaching (Elementary). 6. Technology - Study and teaching (Secondary). I. Zelisko, John, 1950- . II. Title. III. Title: Mathematics, science, and technology connections. IV. Series.

LB1585.C67 1996 372.3'5044 C96-930279-7

Printed and bound in Canada
10 9 8 7 6 5 4 3

This book's text stock contains more than 50% recycled paper.

Safety: The activities in this book are safe when carried out in an organized, structured setting. Please ensure you provide to your students specific information about the safety routines used in your school. It is, of course, critical to assess your students' level of responsibility in determining which materials and tools to allow them to use. As well, please make sure that your students know where all the safety equipment is, and how to use it. If you are not completely familiar with the safety requirements for the use of specialized equipment, please ensure that you consult with specialty teachers before allowing use by students. The publisher and authors can accept no responsibility for any damage caused or sustained by use or misuse of ideas or materials mentioned in this book.

Special Note: This resource has been reviewed for bias and stereotyping.

Editor: Trudy Rising
Project coordinator, proofreader: Diane Klim
Cover designer, Fizzz Design Inc.
production coordinator: Francine Geraci

Acknowledgments

We wish to acknowledge the following Mathematics, Science, and Technology teachers who contributed to this learning resource (school affiliations are those where each teacher taught when this resource was written).

Project Co-ordinator	Bob Corney, Instructional Programs Department, Peel Board of Education
Project Manager/Writer	John Zelisko, Alloa P.S.

Writing Teams

Team 1	Tony McNamara (leader)	Lorne Park S.S.
	Dave Bekkers	Green Glade Sr. P.S.
	Irene McEvoy	Streetsville S.S.
	Stacey Osborne-Howie	Edenwood Sr. P.S.
Team 2	George Gastle (leader)	Fairwind Sr. P.S.
	Rosalind Grant	Homelands Sr. P.S.
	John Sylvester	Gordon Graydon S.S.
	Jim Vincent	North Peel S.S.
Team 3	Bert Stinson (Leader)	Lancaster Sr. P.S.
	John Rodger	Morning Star S.S.
	Sandy Kissoon-Singh	Fletcher's Creek Sr. P.S.
	Ralph Trewartha	Bramalea S.S.
Team 4	Brian Juby (Leader)	Chinguacousy S.S.
	Linda Bowen	Alloa P.S.
	Marg Garrett	Herb Campbell P.S.
	Glen Mauer	Chinguacousy S.S.

Graphics Robert Zelisko

Special thanks is also given to Helen Baker, Dr. Alex Norrie, and Allan Smith for their contributions and to Muriel Karadi for her secretarial support in inputting, formatting, and revising this resource. Finally, we thank Ms. Julie Czerneda for bringing our work to our publisher's attention, and we thank our publisher, Trifolium Books Inc., for publishing this resource, thus making it available to teachers and students everywhere. We are pleased to share our teachers' ideas with others.

~ Chris Bridge, Superintendent, Program Services
Peel Board of Education

Contents

Introduction

Mathematics, Science, & Technology Connections results from the efforts of grades 7, 8, and 9 teachers of mathematics, science, and technology, working together, thereby using the expertise of each as a resource for the others. Our mandate was to develop broad, integrated program areas replacing the focus of subjects and disciplines as separate and distinct. The resource promotes an integrated curriculum to assist students in seeing the links between different subject areas and in a larger context, the direct relevance of what they are learning in school to their needs later in life.

The Challenges we produced and the format we used are provided as a guide for other teachers who may be looking for successful ways to work together to develop integrated activities to enhance their curricula. We wish to stress that we recognize our approach is only one of many ways that could be used to develop such activities. We hope that our approach is useful to other teachers as they are developing their own ideas.

You will find that this book uses a problem-solving approach that is "technological" in nature. That is, the book provides activities in which students are challenged to design, construct, and evaluate their solutions to problems. As they work on their Challenges, they will develop new skills, as they connect learnings from traditionally separate disciplines.

Please note that the activities presented are not intended to satisfy all the stated outcomes for any single course or cluster of courses. Teachers will wish to pick and choose Challenges we have provided to ensure each will increase levels of understanding and competence of their students. It is, of course, important that a Challenge not be used if students have already mastered all the skills and knowledge that the Challenge has been developed to satisfy.

We hope that you find the model we have developed, and the Challenges we have provided for our students useful in your own program.

Rationale

No one can deny that Mathematics, Science, and Technology are closely related. A single problem often has aspects of all three. The aim of science is to understand the natural world, and the aim of technology is to make modifications in the world to meet human needs. Mathematics is used as a tool in answering many of the questions of both science and technology — some would say that mathematics is the "language" of science and technology. Design is critical to technology in the same way that inquiry (active exploration) is critical to science. The problem-solving activities in this book use both the design process of technology and the scientific process of inquiry as is appropriate to the situation/question.

We have focused on broad topics (called *Themes* in this book) as a means of integrating Mathematics, Science, and Technology. The Challenges within each theme build on curriculum skills and knowledge the students are gaining in their everyday courses of study, and can be used as springboards to new topics. Some teachers will wish to use Challenges in helping students construct for themselves what they already know, and what they need to know. We hope you will use our Challenges as you find them most useful in your own school, and we hope you will get together with other teachers in your area to develop more and even better Challenges than these to meet your own needs. We have also provided the template we used to help you in developing your own Challenges.

Assumptions

Integrated programs suggest a need for schools to find ways to adapt school organization to enable teachers to collaborate frequently. This will allow teachers to be better able to:

- assess the relevance of curriculum directions in meeting student needs;
- reflect on and discuss teaching strategies;
- document successes and find solutions to problems.

The book assumes that science classes stress learning through active exploration. It has been well demonstrated that this is the only way to ensure students acquire an adequate understanding of concepts, as well as facility in the investigative techniques of science.

The book assumes that mathematics classes give students the opportunity to make numerical measurements, make scale drawings, use formulas to predict their outcome, explore mathematical relationships through the use of concrete materials, measure the data obtained from their explorations, and compare it to the predicted result (based upon the mathematical model).

In delivering the technological components of the book, it is assumed that most secondary schools have technological education facilities, some junior high schools and elementary schools have technology centers, as well, while others may deliver the components in a regular classroom that is appropriately resourced with work stations equipped with basic tools and materials. Safety should be taken into account by the teacher for every Challenge used.

The Challenges we have provided will extend and enrich students' understanding of mathematics, science, and technology, but we wish to stress that the necessary skills and knowledge from your regular programs will need to be in place for that to happen. Successful integration of mathematics, science, and technology requires that students have enough grounding in each of those pursuits that they can make the meaningful connections that will enhance their understanding of the world.

Getting Started

Mathematics, Science, & Technology Connections contains 24 Mathematics, Science, and Technology activities, called Challenges, all developed using the same problem-solving approach. The book results from the team effort of math, science, and technology education teachers, working together to develop fully integrated activities that meet required learning outcomes. Before checking through the Challenges to decide which ones to use in your own curriculum, please note the following:

How to Use Each Challenge (pages 6-9) provides a breakdown of the format we used to produce all the Challenges in this book. You will see that the Challenges fit under four core themes — *Structures and Mechanisms, Movement, Forces, Environment* (several of the *Environment* Challenges would fit well within a unit on *Characteristics of Living Things, Life Functions*, and *Simple Machines*). We started with many more themes, but narrowed the focus of our sample Challenges to fit under just these four major curriculum areas, providing a model for teachers in other jurisdictions to work together to develop their own Challenges.

Blank Working Pages (pages 10-13) provides blank forms to produce your own Challenges, using our sample format (and for students to devise their own Challenges!). Our Publisher has granted you permission to reproduce these pages for your own classroom use. You may wish to put these forms on computer disk to modify the format as you prefer.

Learning Outcomes (pages 14-16) lists the general learning outcomes the Challenges were developed to meet.

Assessment and Evaluation Strategies (pages 17-19) describes the differences between assessment and evaluation, and provides ideas for assessment through the use of:

- Information gathering
- Conferencing
- Self, peer, and group evaluation
- Student performance profiles

Levels of Achievement (pages 18-19) provides examples of levels of achievement you may wish to consider using in assessing your own students. A sample *Student Performance Profile* is provided which you may feel free to reproduce for your own personal use.

Questions to Ask at the Planning Stage (pages 21-26) provides the questions we recommend you ask yourself before you assign any of the Challenges, as a teaching strategy. Beginning on page 27, we have provided a sample teacher response to these questions for the Challenge entitled "Pass the Buck(et)." (This Challenge can be found on pages 27-30).

Challenges (pages 31-135) provide the 24 Challenges in the four core areas indicated above.

Some Useful Resources (page 139-143) lists books, magazines, CD-ROMs, and videos that you may find useful.

Useful Internet Addresses are listed on pages 144-145.

Sample Suppliers (pages 146-148) lists various building materials, in addition to those materials that can be easily obtained locally.

How to Use Each Challenge

Each of the Challenges outlined on pages 32-135 has been developed using the same problem-solving model. (Blank Working Pages, pages 10-13, is provided for you to produce your own Challenges using the same format.) Please note that we have not designated a grade level for the Challenges. We feel that each can be done by students at grades 7, 8, or 9 (and probably quite effectively at grade 6 and grade 10), with modifications/ extensions as you see fit to suit your student population.

Theme

Many common themes (or broad topics) can be used to develop integrative programs in Mathematics, Science, and Technology. As previously stated, we have chosen four themes that we felt could be easily woven into existing curricula to enhance program delivery. The criteria we used in selecting both Themes and the Challenges to fit within each Theme is that (1) they must have obvious subject connections among the disciplines, that can be readily shown to students and (2) the Challenges must be made "real" to students. The Challenges in this book fit within the following Themes:

- Structures and Mechanisms
- Movement
- Force
- Environment

Title

Each Challenge has been given a title, usually one that will be a "catchy" identifier. You may wish to have students identify their own creative titles, particularly after they have done the first few.

Integration Connections

For your interest, we here refer you to the page on which you will find the integration connections that we have considered as we developed the Challenge. Before beginning the Challenge, as a part of your planning, you will also wish to ensure that you list at least two to three learning outcomes that you specifically expect to be met by students' accomplishing the Challenge.

Situation

Challenges should be introduced in the context of a "real world" situation for students. The situation provides the context for the Challenge students will solve. Often referred to as a "Design Brief," this is an overview of the situation that sets the stage for the Challenge.

Challenge
The Challenge is a general statement of the task that the students will explore, design, make, and evaluate. The Challenge should not be so specific as to suggest there is only one solution. Rather, the Challenge should provide for several options with students being given opportunities to brainstorm and select the best solution to the Challenge.

Possible Materials and Equipment
Before students begin the Challenge, teachers should discuss what materials and equipment are available for students' use. Ideas from students for other materials they think might be useful should be encouraged during this discussion, and you may well wish to add to or completely change the Materials and Equipment listed, based on this discussion.

Parameters
This section identifies the constraints within which students must work to solve the Challenge. Factors to consider include time, space considerations, design team membership, your budget, etc. We have set out parameters that work well for each Challenge but, again, you may wish to change these slightly for your own situation.

Exploring Ideas
In using an open-ended problem-solving model, students should be encouraged to generate a number of ideas for approaches to solve the Challenge. Together, assess the alternatives generated by critiquing them within the established parameters. During this design phase, teachers will need to consider research opportunities, sessions for group problem-solving, and at every opportunity, identify the connections among the Mathematics, Science, and Technological concepts that relate to the Challenge.

NOTE: Before students begin the Challenge, teachers will need to identify and clearly articulate to students the criteria that will be used to assess the process and the product.

Choosing and Building the Solution
Once ideas have been generated, each team's research and collective decision-making should lead them to a selection of their best plan for solving the Challenge. Then they're ready to go!

Reflections

On completion of the Challenge, students need to reflect on the process used, and their solution (product). Students might evaluate the process and their product by considering the following questions:

- Does the solution satisfy the Challenge?
- How could the process used to achieve the solution be improved?
- Did the process provide for shared responsibilities?
- Are there ways that we could have made the solution (product) better?

NOTE: At this point, as well, teachers will assess the process and the product, using the ideas we have provided in the section entitled Assessment and Evaluation, or their own assessment tools.

Extensions

Depending on the grade and capabilities of your students, the Challenges can be enhanced or extended to include:

- a modification to make the design more functional or to do other tasks;
- the use of a computer to assist in the design or to interface and operate the project;
- development of a video of the process and operation of the project;
- development of an instruction manual or technical report on "how to build" the project;
- a discussion of career options associated with the Challenges, and the types of skills and interests one would need to consider the career;
- a discussion of possible impacts the team's solution to the Challenge might have on the environment (either positive or negative).

Some Useful Resources

Although students should be encourage to, and often will find their own resources appropriate to the tasks, teachers are advised to list those resources that should be referenced as a starting point.

Integration Connections

This section lists *Mathematics, Science, & Technology Connections* to the Challenge. In most cases, these connections will represent prerequisite knowledge needed to understand fully the applications to the Challenge. As noted in the Introduction, it is essential that this knowledge be acquired in a way that ensures genuine understanding of the concepts and also develops competence in the processes and techniques specific to each subject discipline.

Teacher Talk

The *Teacher Talk* section for a Challenge may take one of several forms. It may:

- sketch a possible solution to the Challenge;
- sketch how science experiments and mathematical concepts can be related to the project;
- provide reminders of teaching strategies that might assist with the implementation of the project.

NOTE: It is important that teachers not allow the *Teacher Talk* sections we have provided as our own suggestions to influence students' solutions to the Challenges. If students see the possible solutions we have provided before they have an opportunity to generate their own ideas, they may be influenced by our suggestions and their own creativity may thus be stifled. Most Challenges if properly designed and introduced will allow for many possible solutions and satisfy the outcomes appropriate to each subject discipline.

Blank Working Pages

Theme:

Title:

Integration Connections:

Situation:

Challenge:

Materials & Equipment:

Parameters:

Exploring Ideas:

Choosing & Building the Solution:

Reflections:

Extensions:

Some Useful Resources:

Integration Connections

Mathematics	Science	Technology

Teacher Talk:

Learning Outcomes

As we move into the 21st Century, education is undergoing dramatic change. One such change is that most curricula now focus on results — or "learning outcomes" — rather than on objectives or goals. The Learning Outcomes listed below are general ones that the writers used in formulating the Challenges in this resource. We feel confident that teachers everywhere will easily be able to use the Challenges to address specific Mathematics, Science, and Technology Outcomes required in their own state or province.

Inquiry

- develop and carry out plans that combine appropriate strategies to obtain and record relevant information in appropriate formats; try another approach if necessary
- decide when they have enough useful information; select and organize the information to reveal meaning; make links to other knowledge
- demonstrate or describe in sufficient detail what was done and what was learned or produced; explain and support conclusions and points of view

Problem Solving

- develop and carry out plans to solve problems, consider different approaches, break complex problems into sub-problems; find necessary information; recognize when an approach isn't working and try another
- demonstrate or describe what they did to solve problems and evaluate the results and the chosen strategies using their own and given criteria.

Technological Competence

- select and use appropriately the most effective combination of technologies, tools and instruments to: observe and measure; manipulate and alter materials; create products
- use a variety of technologies to obtain, record, organize, create, and present information
- assess the process and results of their own use of technologies for appropriateness, effectiveness, and safety

Self Management

- express ideas and feelings in ways that are appropriate to the situation

- develop and use a variety of strategies to initiate activities, meet requirements, adapt to new situations and make improvements in behavior and learning

Interdependence

- interact positively with others by:
  - sharing space and resources
  - giving and accepting help and encouragement
  - respecting the rights of others

- advance a group's task by:
  - contributing ideas
  - building on the ideas of others
  - dividing tasks equitably
  - accepting and fulfilling responsibilities
  - monitoring progress
  - adjusting roles and tasks as required

- develop and use their own and given criteria to assess a group's process and products, and their own performance

Acquiring Information

- choose and locate appropriate sources of information in their school, community and beyond

- make accurate, detailed observations, selecting the appropriate aids for observation and measurement

Processing Information

- summarize details, main ideas and concepts, draw conclusions and make supported judgments

- use estimation, computation and measurement to determine quantities, to clarify relationships and to make and confirm predictions

Organizing Information

- describe, explain, and compare objects, events, and ideas

- sequence a group of objects/events, ideas according to criteria chosen to suit the purpose

Using Conventions

- use appropriate vocabulary including subject specific words and symbols to express and clarify meaning, and give instructions

- follow or develop rules, instructions, and procedures to complete tasks and improve performance or product

Communication

- communicate effectively in informal and formal situations in ways that are appropriate to the purpose, information, audience, and context

- communicate and create products for different purposes including:
  interacting with others
  expressing and elaborating on personal feelings, ideas, and experiences
  giving an organized, comprehensive account of facts and ideas
  expressing opinions, values, and arguments supported by evidence
  constructing representations of ideas, concepts, and procedures

Assessment and Evaluation Strategies

Information Gathering
One of the best ways to observe and collect information on students' daily work is through the use of Student Portfolios. Sometimes referred to as Student Journals or Design Folders, a Portfolio is a collection of student work, and provides opportunities for students to document their own progress over a long period of time. Portfolios provide:

- a collection of daily work that will identify student areas of strength and weakness. These strengths or weaknesses are visual records that can provide the basis of student/teacher consultations and for parent discussions;
- immediate evidence for students and teachers where remediation is required or program modification advised;
- a shared evaluation when a Challenge for students is integrated in the Mathematics, Science, and Technology areas. Over time, students will see the relevance between subject disciplines, and teachers will be less inclined to departmentalize concepts and skills.

When Portfolios are introduced to the class, students must be clearly aware of the knowledge, skills, and attitudes areas that will be observed while engaged in individual and group activities. These components may be appended to the portfolio with opportunities for teachers to provide comments on observations made during student activities. Providing opportunities for students to include their own comments will provide the basis for a conference situation to decide where marks for the identified skills might be placed.

Conferencing
Recent educational policies promote the philosophy of teachers and students working in consultation with each other about student learning. The materials assembled in portfolios is one source of providing the basis for consultation. Conferences that are well planned with openness, constructive criticism, and clearly identified student expectations as key elements, assist students in receiving immediate feedback as work is submitted and to understand and/or modify their own strengths and weaknesses.

Self, Peer, and Group Evaluation
When collaborative groups of students are assembled to collectively solve a Challenge, each student should be helped to understand the value of

assessing personal performance. An atmosphere of trust and openness should be modeled by the teacher in student/teacher conferencing. This can provide the foundation for developing a similar atmosphere of trust and honesty in assessing their own and their peers' performance. Teachers must also be willing to accept and place a value on student contributions to the assessment and evaluation process.

Student Performance Profile
To assist teachers in identifying the levels of achievement reached by a student as he/she works towards achieving the identified outcomes for the Challenge, individual student performance profiles may be developed. These profiles should list all the knowledge, skills, and values, at several levels of achievement for each Challenge. These profiles could also be part of a student portfolio that will provide opportunities, for all teachers and students involved, to place their comments appropriate to the outcome statement. It will also give students, teachers, and parents a "snap-shot" of where students are in achieving the outcomes. With the integrated nature of the program, the profile should be developed in consultation with each of the teachers involved to:

- include the appropriate outcomes to reflect the subject specific and integrated components of the Challenge or program;
- develop a range of achievement levels that are appropriate for each teacher/program.

Levels of Achievement
Several schools/boards, and subject disciplines are developing descriptors of achievement levels. The following are offered as an example of descriptors that could be used:

Awareness Level
Students are aware of the process to achieve a specific outcome but have not yet applied the process. Through personal observation, they would have been exposed to the specific outcome.

Application Level
Students have obtained some practise in the process or activity to achieve the outcome but have not yet achieved a skill level. The practise would have been achieved only with the support and guidance of a facilitator or with resource support.

Competent Level
Students have achieved a level of skill in the process or activity that consistently demonstrates the attainment of the specific outcome. The level of skill would have been demonstrated without additional support.

Mentor Level

Students have attained a level of expertise with the specific outcome that demonstrates their ability in a leadership or consultative role. The student would understand the process well enough to analyze, teach, and perhaps improve upon the processes described in the outcome statement.

The following profile is a sample that could be used to identify levels of achievement for student performance in problem solving and technological competence (see Learning Outcomes described on pages 14-16). Please note that this sample profile could be modified in a number of ways to fit your own needs.

Student Performance Profile

Program: ________________ Student: ____________________ Year: _____

Outcome Type (e.g., knowledge, skill, etc.)	Performance Indicators (demonstrations)	Awareness	Application	Competent	Mentor	Date and/or Comments
	The student has: Problem-solving • developed and carried out plans to solve problems. • considered different approaches to solve problems. • recognized when an approach to solve a problem does not/will not work. Technological Competence • selected and used appropriate combinations of tasks and technologies to create a product. • used a variety of technologies to obtain, record, organize, create, and present information. • assessed the process and results of the student's own use of technologies for appropriateness, effectiveness, and safety.					

Questions to Ask at the Planning Stage

1. For this theme (topic), which Challenge would best meet the needs and interests of the students and the available resources at this time?

2. Which cross-curricular Learning Outcomes will be a focus?

3. What concepts and skills do we expect students to learn (or confirm they know) as a result of this Challenge

in Mathematics?

in Science?

in Technology?

4. What understandings and skills need to be in place prior to presenting the Challenge in Mathematics?

in Science?

in Technology?

5. How will the Challenge be used
a) as a way of helping students to construct what they know?
b) as an introduction to the theme/unit/concept?
c) as a vehicle for teaching concepts?
d) as a follow-up to other learning?

6. What changes in wording or design parameters are necessary to make the Challenge better at incorporating the specific Mathematics, Science, and/or Technology learning which is to be a focus?

7. What opportunities will be available during and after completion of the Challenge to identify and test variables to improve performance?

8. How will we evaluate student learning?

QUESTIONS TO ASK AT THE PLANNING STAGE

(Please see the *Sample Challenge* on pages 27-30.)

1. For this theme (topic) which Challenge would best meet the needs and interests of the students and the available resources at this time?

   Theme: *Force*

   Title: *"Pass the Buck (et)"*

   Challenge: *a model of a water wheel to harness the motion of water and to drive a millstone.*

2. Which cross-curricular Learning Outcomes will be a focus?

   1. *Interdependence*
   2. *Problem Solving*
   3. *Technological Competence*

3. What concepts and skills do we expect students to learn (or confirm they know) as a result of this Challenge:

Mathematics

- *finding the center of a circle. e.g., using a compass, paper folding*
- *measuring angles using a protractor*
- *discovering (or verifying) that the sum of the angles in a n-gon is 180° (n-2).*
- *constructing a circle with a given ratio and determining where the vertices for an n-gon should go*
- *investigating the effects of different gear ratios.*
- *applying the Pythagorean Theorem in designing the building supports.*
- *calculating the volume of 3-D figures*
- *classifying a triangle as equilateral, isosceles or scalene*
- *graphing the relationship between two variables: e.g., capacity of a container OR radius of wheel OR number of sides of wheel VS. the energy created (refer also to Science concepts and skills)*

Science

- *hypothesizing the effect of altering a particular variable.*
- *fair testing — explore the effects of altering one variable while controlling other variables: e.g. changing radius of water wheel, number of sides of second wheel, number or paddles/scoops on water wheel, height of water on wheel, materials used for paddles.*

NOTE: *The energy of the millstone can be measured by counting the number of rotations per minute. This "measurement" is adequate for comparison purposes.*

Technology
- *describing the components, roles, and responsibilities within the problem-solving process.*
- *identifying and demonstrating the safe use of tools and equipment appropriate for the Challenge*
- *selecting, preparing, and assembling materials*
- *building prototypes of mechanisms linked to a water wheel and determining the efficiency of rotating parts.*

4. What understandings and skills need to be in place prior to presenting the Challenge?

Mathematics: *Students will have learned how to find the center of a circle.*

Science: *Students will have been introduced to the concept of fair testing.*

Technology: *Students will have been introduced to the use of a water wheel and its mechanism(s).*

5. How will the Challenge be used?
   - *as a vehicle for applying mathematical knowledge about circles*
   - *as a means of graphing relationships between variables*
   - *as a means to develop the processes of hypothesizing and fair testing*
   - *as a means to experience a problem-solving process*
   - *as a means to select appropriate materials, assemble combinations of materials and mechanisms, and evaluate the efficiency of moving parts.*

6. What changes in wording or design parameters are necessary to make the Challenge better at incorporating the specific Mathematics, Science, and/or Technology learning which is to be a focus?

be more specific within the parameters to ensure there is consistency with whole-class comparison of results. e.g. by adding a parameter that the millstone should have a diameter of 10 cm will ensure consistency when a class is measuring "energy" by counting the number of rotations per minute

7. What opportunities will be available during and after completion of the Challenge to identify and test variables to improve performance?
   - *accuracy required in finding the center of a circle*
   - *fair testing procedures*
   - *appropriate selection and assembly of materials*

8. How will we evaluate student learning?

Observation

- *variables are controlled in testing*
- *significant problems in the use of materials and construction procedures are dealt with and not ignored*

Performance Assessment

- *given a circle with a diameter of 30 cm, students will:*
  find the center
  turn the circle into an equilatoral, 8-sided figure

Self Assessment

- *what I did well*
- *what I had trouble with*
- *what I would do differently next time*

Special Planning Notes

Theme:

FORCE

Title:

"Pass the Buck(et)"

Integration Connections:

See page 30.

Situation:

The early pioneer farm was a difficult place to work. Animals and children were used to doing much of the work on the farm including the grinding of grain to make flour.

Challenge:

Design and build a working model of a water wheel that will harness the motion of moving water or the gravitational force of falling water to turn the grinding stone in a mill.

Materials & Equipment:

Wood: basswood or willow strips 1 cm x 1 cm x 1 m and 0.5 x 1 cm x 1 m in length; 1/16"
1/4" dowel; Bamboo BBQ skewers; round wooden disks: 2-3" diameter
Tools: hack saw, miter box or miter saw, assorted drill bits, glue gun, 30 cm ruler, hand drill
Drafting materials: set squares, architects scale, T square, protractor, compass
Other: Lego 1030 Kits; plastic cups (35 mm film containers or substitute) worm gears
optional: hole saw, drill press, scroll saw, sanding drums if disks are to be manufactured

Parameters:

A maximum of 4 m of wood strip materials; maximum size of water wheel 30 cm x 30 cm.
Students are to work in groups of 3.
The device should be of durable construction.
The water wheel must rotate freely with minimum friction.
The water wheel must turn a gear to change vertical motion into horizontal motion.
Time frame 6 hours of class time.

Exploring Ideas:

Students will:
Research the design and purpose of historic water wheels;
Examine the positioning and operation of a water wheel in a river or water fall;
Examine the use of gears and different gear configurations to regulate speed and change motion direction;
Experiment and safely manipulate tools and materials in the most efficient manner.

Choosing & Building the Solution:

Students will: Brainstorm ideas, produce rough sketches, negotiate the best solution, create a model or prototype, revise the model, select the final design, create working drawings, and construct the final machine.

Reflections:

Were all the safety considerations recognized?
Was the motion of the water harnessed by the water wheel and transferred into horizontal motion?
Was the construction technique successful?
What made a successful water wheel?
What other machines, materials, tools would have made the Challenge easier to complete?
What human (group dynamics) problems were encountered?
If done again, what would you have changed/do differently?

Extensions:

Examine rotation of a device using water power and apply principles to the design and construction of a wind machine.
Change size and shape of the water wheel (mullet-sided).
Change size (diameter) of the gears; vary number of sprockets or teeth in the gears
- experiment with different gear constructions
- harness the created energy to do other "work" activities.

Draw a scale orthographic drawing of the water wheel.

Some Useful Resources:

Macaulay, David, *The Way Things Work*.
The Metropolitan Toronto School Board, *By Design: Technology Exploration & Integration*.
Czerneda, Julie, *Great Careers for People Who Like to Work with Their Hands*.
Richardson, Peter & Richardson, Bob, *Great Careers for People Interested in Math & Computers*.
Richardson, Peter & Richardson, Bob, *Great Careers for People Interested in How Things Work*.
Any high school physics textbook
Lego Technic Kits (see Sample Suppliers).
Students may wish to check useful WWW pages (see Useful Internet Addresses)

Integration Connections

Mathematics	Science	Technology
• find the center of a circle • measure angles • sum of angles in an n-gon = 180 (n-2) • construct a circle with a given radius & determine where the vertices for an N-gon should go • gear ratios • Pythagorean theorem (building supports) • volume of the cylindrical cups • vary size of containers and compare to the energy generated (graph info) • how does the number of sides affect the energy? • unit rate = number of revolutions/minute (or second) • vary radius & compare energy generated	• gravitational potential, energy- converted to kinetic energy • definitions & understanding of simple machines - wheel & axle, gears changing direction of force, making work easier • Newton's laws of Physics • observe RPM's of water wheel • measure force of water turning the wheel (attach spring scale to a downward bucket & measure the resistance (force required to stop the movement) • volume of buckets — measure — set up controlled experiment to vary this & observe affect on RPM's	• research rotational motion in the water wheels • safe use of tools and equipment • selection of appropriate materials and combinations of materials • problem solving process • brainstorming • sketching • prototype making • preparation, shaping, and processing materials • assembly of materials — drilling, joining, fastening, assembly of gears and supporting structure • appropriate use of materials and the best construction • efficiency-mechanical efficiency of the machine gears, low friction; rotating of the wheel (strength of construction) • group dynamics — what problems were encountered in group assignment working together? • include a written report

Teacher Talk:

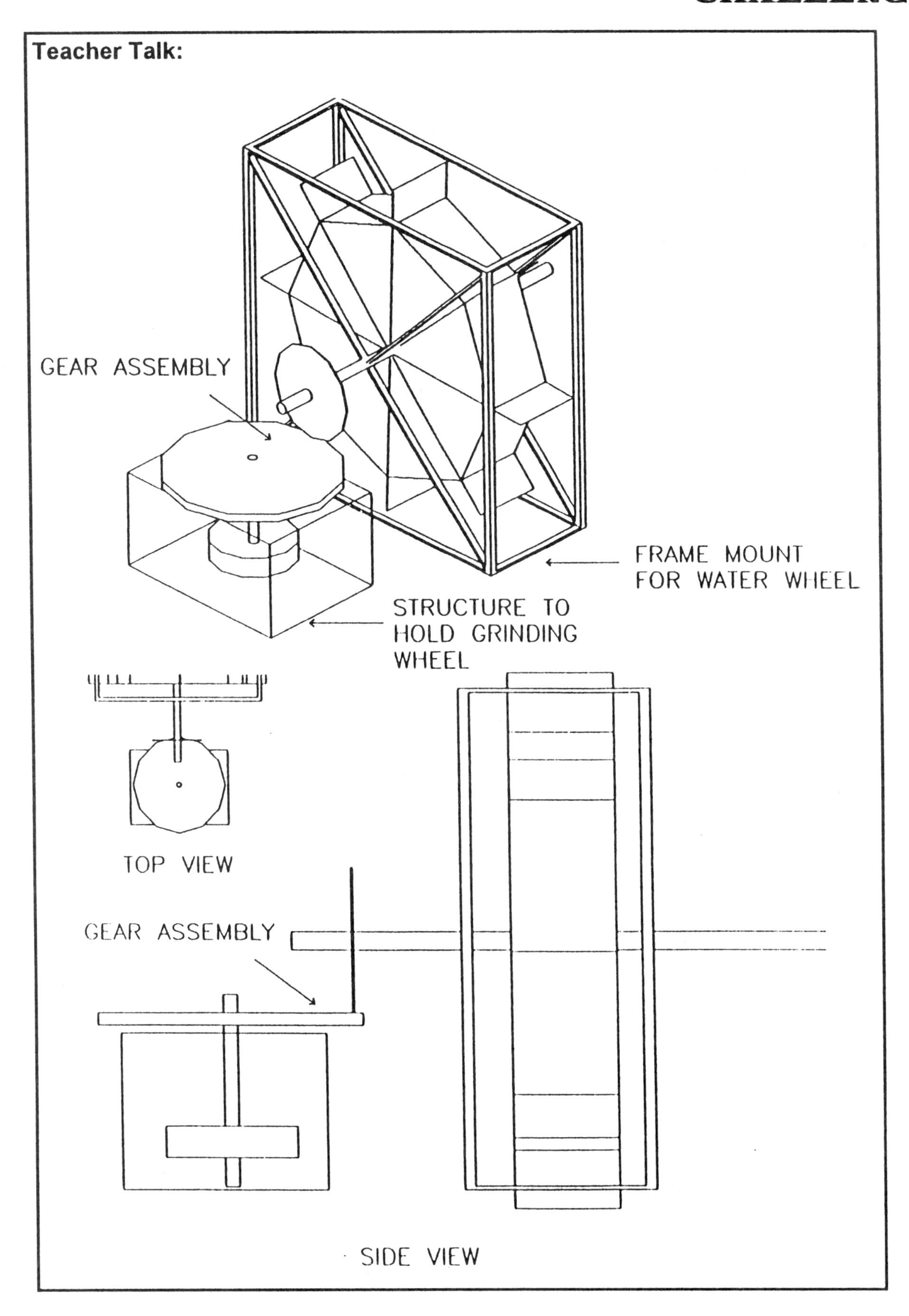

Structures and Mechanisms Challenges

Introducing Challenges

Before you begin this section, please consider *Questions to Ask at the Planning Stage* on pages 21-23. Space is left below for your notes on how you might change your strategy next time or perhaps to record in which classes you have used each Challenge.

Teacher Notes

Theme:

STRUCTURES AND MECHANISMS

Title:

"When the Going Gets Tough" (Wheels & Axles)

Integration Connections:

See page 36.

Situation:

You have been contracted by Spar Aerospace to design a mars vehicle to carry supplies for the astronauts over its tough terrain.

Challenge:

Design and construct a prototype self-propelled vehicle called a Mars Rover capable of traveling 2 m over a sand base containing rocks up to 5 cm in diameter and to carry a minimum 1 kg of mass.

Materials & Equipment:

Could include:

1 cm x 1 cm wood pieces, Bristol board triangles, (Jinks method), wood glue, dowel, steel rod, corrugated cardboard, rubber bands, straws, tongue depressors, recyclable materials, scissors, masking tape, miter saw, square, drill, drill bits, soldering guns, solder, wires, electric motor (9 V or less), batteries.

Parameters:

Vehicle length maximum of 20 cm.
Maximum weight of 1.5 kg empty.
Must travel on its own power.
If operated electrically, must be 9 V or less.
Students should work in groups of 3.
Priority given to recyclable materials.

Exploring Ideas:

What type of terrain can be found on Mars?
What type of power source will we use?
What should the chassis look like?
What type of rolling mechanism will allow the vehicle to move?

Choosing & Building the Solution:

Students will select their best solution, complete working drawings, construct the Mars Rover, and revise on an on-going basis with emphasis on safety and readily available recyclable materials.

Reflections:

Did the Mars Rover meet the conditions in the Challenge?
What problems occurred in the process?
Did the group members share responsibilities?
What other resources would have made the Challenge easier or cheaper to complete?
If you were to do it again, what would you do differently?

Extensions:

Cost out the Mars Rover for cost of materials.
Investigate other applications for a vehicle such as this one.
Research how it might be transported to Mars with respect to portability, type of materials, payload, time travel, etc.
What careers could you associate with this Challenge? Investigate them further.
Create working drawings using a computer drawing program (e.g., Auto Sketch).

Some Useful Resources:

Kellett, Joe & Jinks, David, *Design and Make Activities Series*, DJK Technology, 1993.
The Metropolitan Toronto School Board, *By Design: Technology Exploration & Integration.*
Czerneda, Julie, *Great Careers for People Who Like to Work with Their Hands.*
Richardson, Peter & Richardson, Bob, *Great Careers for People Interested in How Things Work.*
Resource center for space related materials.
Holland, Peter, *Amazing Models: Rubber-band Power*, Tab Books Inc.
Corney, Bob & Dale, Norman, *Technology I.D.E.A.S.*

Integration Connections

Mathematics	Science	Technology
• measurement • cost analysis (extension) • scale drawings • ratio and proportion • charting and graphing • data gathering • calculation of speed • mass versus time	• definition of force • SI unit of force (N-Newton) • types of forces • properties of materials • electrical circuits • measurement of mass • qualitative power analysis • data collection (distance and time)	• open ended problem solving • materials selection • processing materials • assembly of materials — joining, hinging, fastening • safe use of tools • working drawings and sketches • efficiency of operating mechanisms

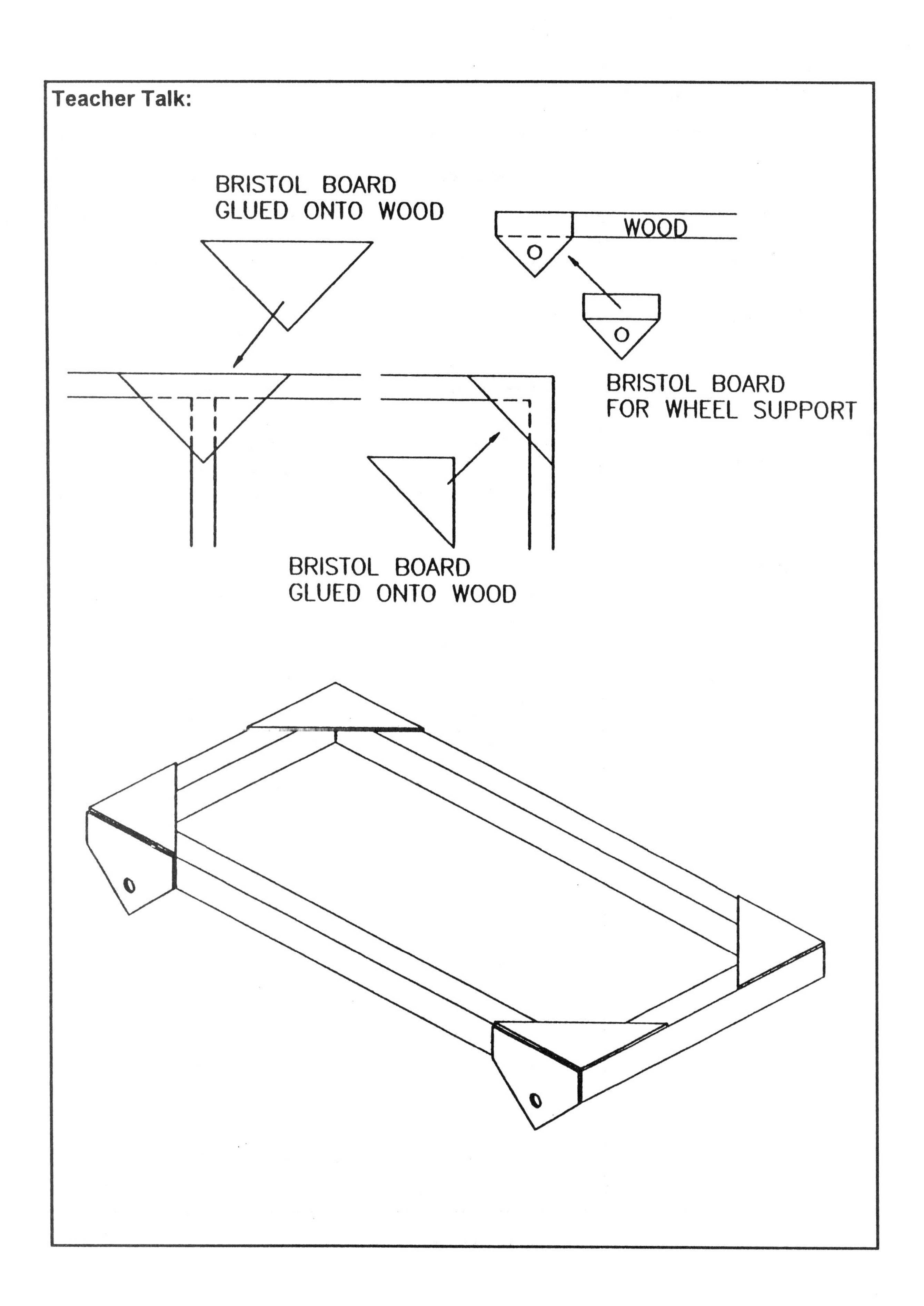
Teacher Talk:
BRISTOL BOARD
GLUED ONTO WOOD
WOOD
BRISTOL BOARD
FOR WHEEL SUPPORT
BRISTOL BOARD
GLUED ONTO WOOD

Theme:

STRUCTURES AND MECHANISMS

Title:

"An Uplifting Experience" (Levers)

Integration Connections: See page 40.

Situation:

You have been hired by MST Fun Fair Inc. to provide a variety of amusements incorporating simple levers.

Challenge:

Design and construct a device that makes use of a simple lever to propel an object (plastic golf ball) to a vertical height of 1 m.

Materials & Equipment:

Could include:
Bristol board, wood glue, 1 kg bag of sand, 2 L (64 oz.) pop bottles, plastic bottles, nails, wire, 1 cm x 3 cm x 50 cm long wood pieces, exacto knife, tape, scissors, ruler, drill, drill bits, saw, plastic practice golf ball, 5 cm x 10 cm x 20 cm long base made from cardboard or plywood to hold the structure.

Parameters:

Maximum of one kg mass weight.
Maximum vertical height of propelled object is one metre.
When optimum height is achieved, machine parts must be made permanent.
Use a plastic practice golf ball.
Students will work in small groups.
Priority given to recyclable materials.

Exploring Ideas:

Explore three classes of levers.
Explore mechanical advantage.
What will the device look like?
Research various amusement devices.
What types of materials could be used?

Choosing & Building the Solution:

Our example:
A "test your strength" machine capable of raising a ball 2 m high using a 1 kg mass source of energy dropped freely from a height of 1 m

Reflections:

Did the device meet the conditions in the Challenge?
What problems occurred in the process?
Did the group members share responsibilities?
What other resources would have made the Challenge easier or cheaper to complete?
If you were to do it again, what would you do differently?

Extensions:

Students are to design their own entrapment device to hold the ball.
Make it more esthetically attractive.
Develop ideas of how to encourage more people to want to try this machine.
Develop other amusement machines using this simple lever mechanism.
Develop a mini MST amusement fair or park and encourage other students to try it.
Create working drawings using a computer drawing program (e.g., Auto Sketch).
Make this machine portable to be taken apart and erected again.
Field trip to an amusement park.

Some Useful Resources:

Kellett, Joe & Jinks, David, *Design and Make Activities Series* [Folios include: Design & Technology in Perspective, Tools & Materials, Box Modelling, Frames and Structures, Making Things Move, Mechanisms I, Mechanisms II, Mechanisms III, Electrics, Computer Control] DJK Technology, 1993.
Local amusement park
Videotape: *Eureka Series Unit 2* Simple Machines.
Corney, Bob & Dale, Norman, *Technology I.D.E.A.S.*
Williams, Peter & Jacobson, Saryl, *Take a TechnoWalk.*
The Metropolitan Toronto School Board, *By Design: Technology Exploration & Integration.*

Integration Connections

Mathematics	Science	Technology
• measurement • use of formula • scale drawings • calculate force x distance • charting and graphing • data gathering • constant and variable measurement	• definition of levers (3) • kinetic & potential energy • mass • gravity • discussion of transfer of energy	• open ended problem solving • materials selection • processing materials • assembly of materials — joining, hinging, fastening • safe use of tools • working drawings and sketches • recyclable materials • efficiency of operating mechanisms

Teacher Talk:

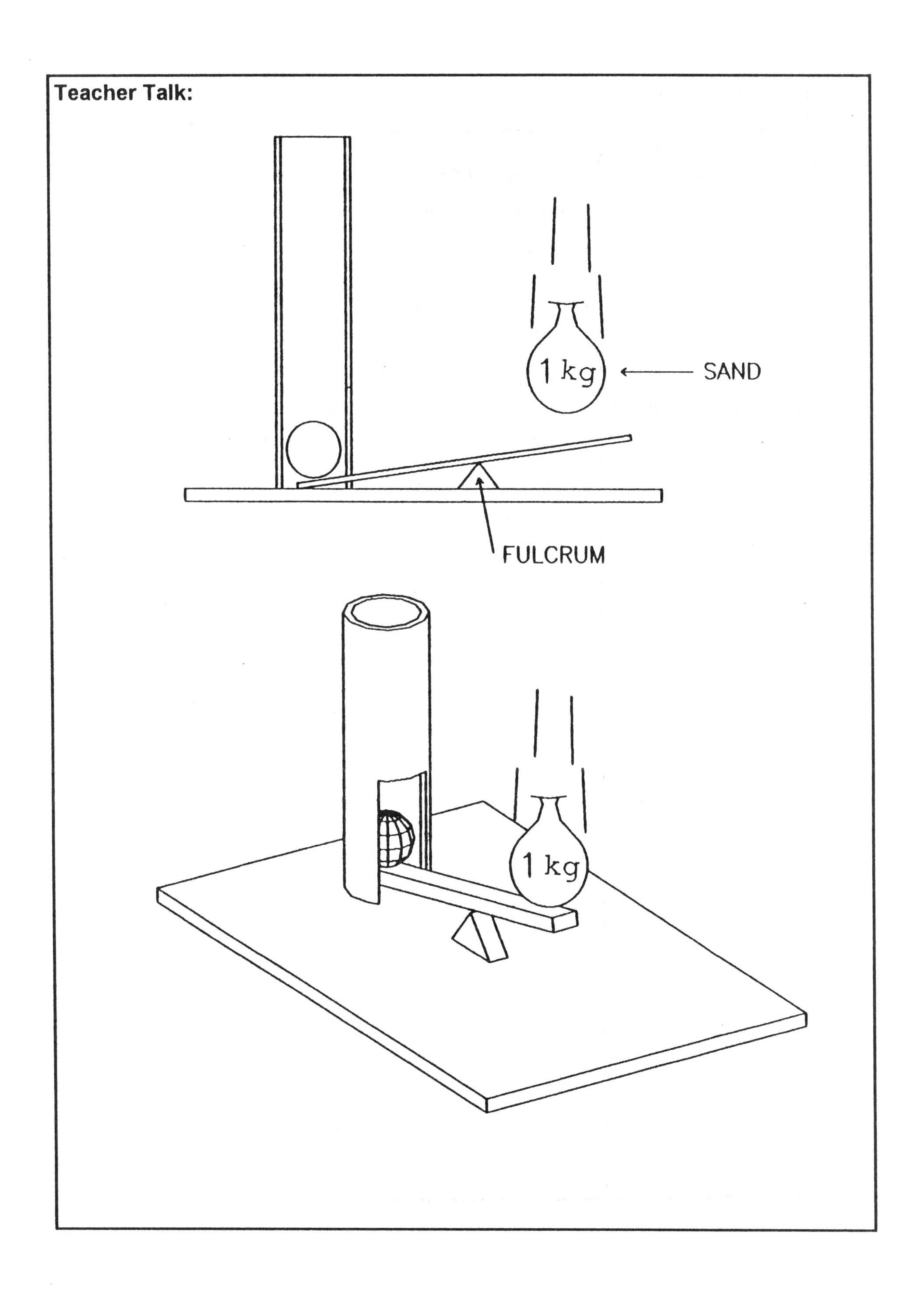

Theme:

STRUCTURES AND MECHANISMS

Title:

"Fun at the Fair" (Gears)

Integration Connections:

See page 44.

Situation:

You have been hired by MST Fun Fair Inc. to provide a variety of amusement rides incorporating gears and/or pulleys.

Challenge:

Design and construct a device that makes use of gears and/or pulleys.

Materials & Equipment:

Bristol board, corrugated cardboard, tongue depressors, coffee stir sticks, scissors, tape, glue, straws, wood pieces 1 cm x 1 cm x 30 cm, elastic bands, string, base to support unit — cardboard or plywood.

Parameters:

Use a minimum of 3 gears and or pulleys.
Circular base for carousel should be a minimum of 20 cm in diameter.
Must have a gear reduction.
Must be firmly mounted on a base.
Esthetically pleasing.
Students will work in small groups of three.

Exploring Ideas:

Explore gear ratios and types of gears and pulleys.
What will the device look like?
Research various amusement rides.
What types of materials can be used?

Choosing & Building the Solution:

Our example:
A carousel that will rotate by means of a crank and gears.

Reflections:

Did the device meet the conditions of the Challenge?
What problems occurred in the process?
Did the group members share responsibilities?
What other resources would have made the Challenge easier or cheaper to complete?
If you were to do it again, what would you do differently?

Extensions:

Field trip to amusement park.
Make it more esthetically attractive.
Develop ideas that would encourage more people to want to try it out.
Use cams for alternative moving devices on the ride.
Specify gear ratios and direction to be used.
Develop a gramophone that will work.
Use computer aided drafting to produce working drawings.
Make the machine portable so it can be disassembled and reassembled.

Some Useful Resources:

Kellett, Joe & Jinks, David, *Design and Make Activities Series*, DJK Technology, 1993.
Local amusement park
Videotape: *Eureka Series Unit 2* Simple Machines, programs 11 to 15.

Integration Connections

Mathematics	Science	Technology
• measurement • scale drawings • use of formula • gear designs • measuring gear ratios • charting and graphing • data gathering	• gears and pulleys • gear ratios and direction • types of gears • pulleys • calculating gear ratios • evaluate choice of pulley and gears as appropriate for the mechanical advantage required	• open ended problem solving process • materials selection • processing materials • assembly of materials — joining, hinging, fastening • safe use of tools • sketches • working drawings • recyclable materials • strength of assembly • efficiency of operating mechanisms • aesthetics • ergonomics

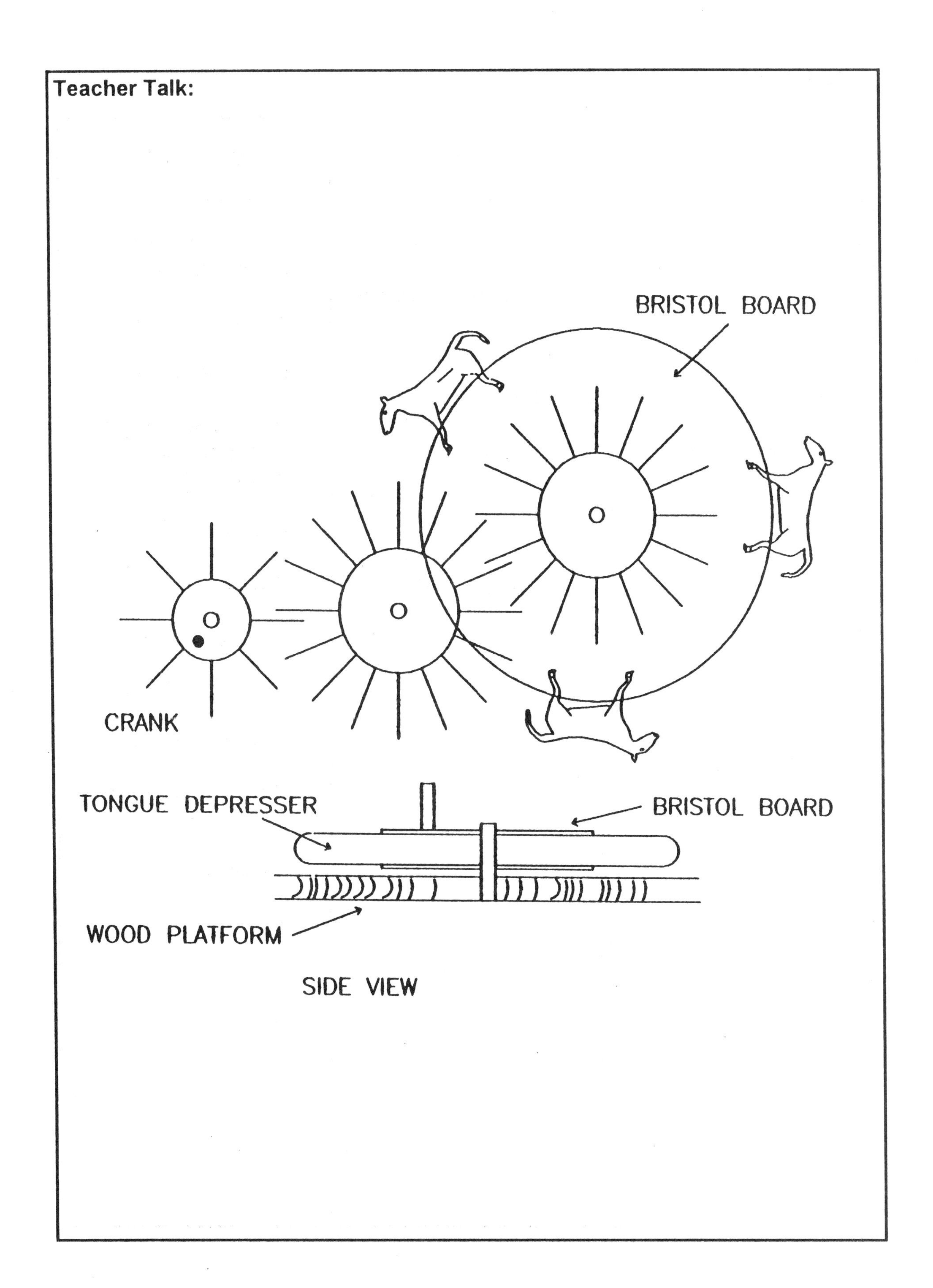
Teacher Talk:
BRISTOL BOARD
CRANK
TONGUE DEPRESSER
BRISTOL BOARD
WOOD PLATFORM
SIDE VIEW

Theme:

STRUCTURES AND MECHANISMS

Title:

"Air Ball" (Hydraulics & Pneumatics)

Integration Connections:

See page 48.

Situation:

Science World has asked you to demonstrate the workings of a pump for one of their upcoming TV shows.

Challenge:

Design and build a pump to create the greatest lift-off of a ping-pong ball.

Materials & Equipment:

scissors, 2 L (64 oz.) pop bottle, ping-pong balls, small recyclable pop bottle, tape, glue, 6 mL plastic inner tube for diaphragm, glue gun, wooden discs (Wheels) for diaphragm, ruler, doweling, threaded rod, exacto knife, Vaseline, grease, soap
OPTIONAL: Use ABS piping and fittings.

Parameters:

One stroke operation (Draws air — on "in" stroke, expels air (ping-pong ball) on "out" stroke).
Students should work in small groups.

Exploring Ideas:

Research pump capacity and volume.
What will the device look like?
Research valves and pressure.
Test types of sealing materials.
Test sizes of openings and cylinders.
Research how a pump works, i.e. take apart a bicycle pump.

Choosing & Building the Solution:

Build a pump using a 2 L pop bottle with an inlet and outlet valve (ping-pong ball launching) with materials supplied.
Build measuring device to measure height.

Reflections:

Did the device meet the conditions of the Challenge?
What problems occurred in the process?
Did the group members share responsibilities?
What other resources would have made the Challenge easier or cheaper to complete?
If you were to do it again, what would you do differently?

Extensions:

Video tape presentation — storyboarding, videotaping, scripting.
Demonstrate to junior classes.
Uses or applications — toy, heart machine, etc.
Use computer aided drafting to produce working drawings.

Some Useful Resources:

Have pumps on hand. Disassemble various types of pumps and study how they work.
Optional: Pneumatics/hydraulics kits from suppliers (see Sample Suppliers)
Discuss the heart as a pump — any science textbook introducing the circulatory system.

Integration Connections

Mathematics	Science	Technology
• measurement • multiplication and division of decimals • scale drawings • use of formula • measurement of volume and area • area calculations • charting and graphing • data gathering	• volume, pressure, and area • force per unit area • viscosity	• open ended problem solving process • materials selection • processing materials • assembly of materials — joining, hinging, fastening • safe use of tools • sketches • working drawings • recyclable materials • strength of assembly • efficiency of operating mechanisms • aesthetics • ergonomics

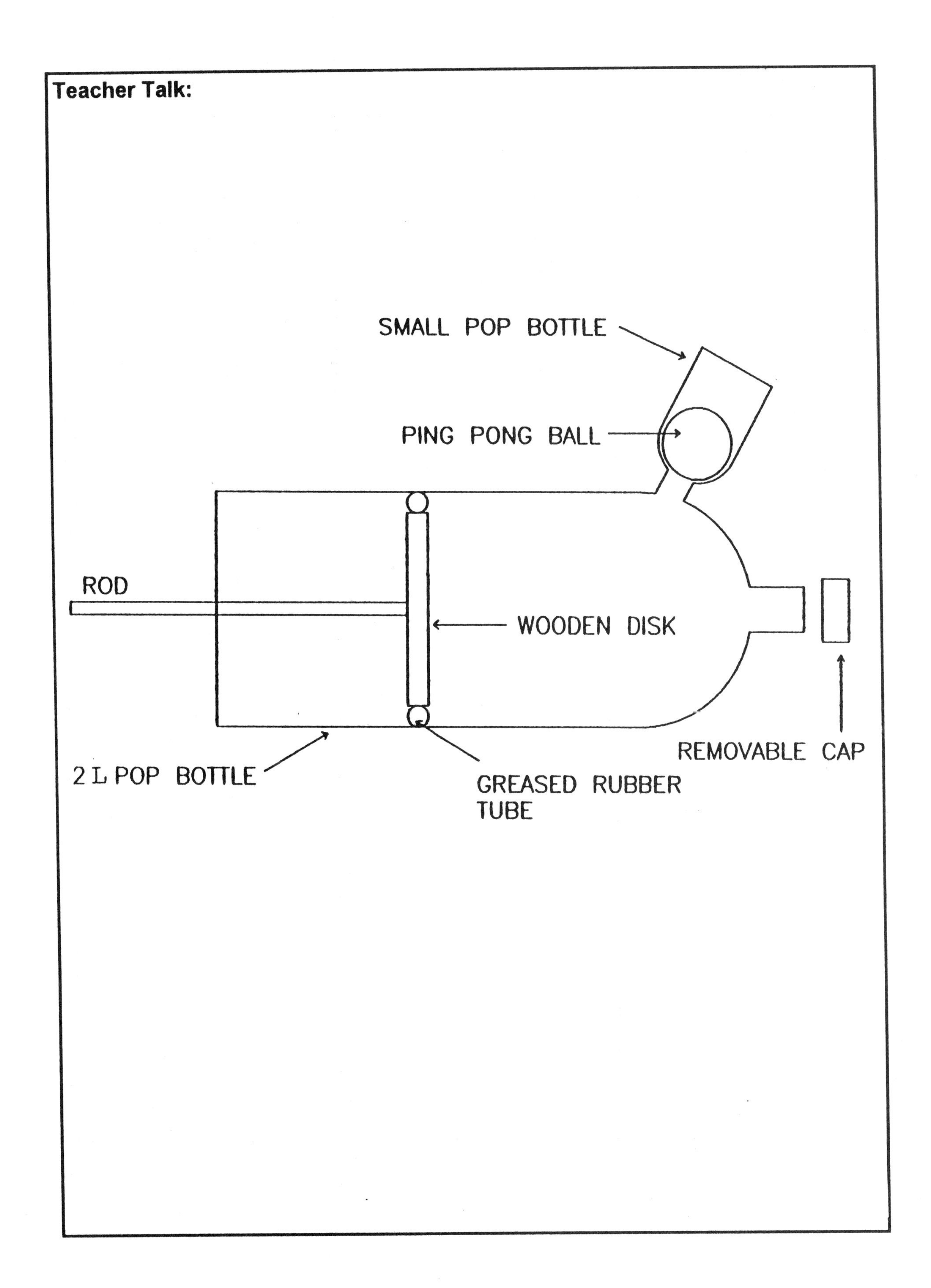
Teacher Talk:
SMALL POP BOTTLE
PING PONG BALL
ROD
WOODEN DISK
REMOVABLE CAP
2 L POP BOTTLE
GREASED RUBBER
TUBE

Theme:

STRUCTURES AND MECHANISMS

Title:

"Why Do It the Easy Way" (Assortment of Mechanisms)

Integration Connections:

See page 52.

Situation:

You have been hired by a MST Cartoon company to develop a machine that demonstrates a difficult and amusing series of mechanisms to perform a single task.

Challenge:

Design and construct an amusing device using a minimum of 3 of the following mechanisms: levers, pulleys, hydraulics, pneumatics, gears, cams, wheels & axles.

Materials & Equipment:

Common items found around the house.
Students are to bring in items.
Plywood for base or backing.

Parameters:

Must fit on student desk.
Must incorporate a minimum of 3 mechanisms.
Could be used as a follow-up project at end of the structures and mechanisms unit.

Exploring Ideas:

What will the device look like?
Group discussions and brainstorming of possible mechanism combinations.
Research these mechanisms.

Choosing & Building the Solution:

Our example:
A machine that will pour coffee into a cup and add a lump of sugar to it.

Reflections:

Did the device meet the conditions of the Challenge?
What problems occurred in the process?
Did the group members share responsibilities?
What other resources would have made the Challenge easier or cheaper to complete?
If you were to do it again, what would you do differently?

Extensions:

Design a device to heat the coffee; add cream to coffee.
Increase the number of minimum devices to be used.
Video tape presentations — storyboarding, videotaping, scripting.
Demonstrate to junior classes.
The wackier the better!
Use computer aided drafting to produce working drawings.

Some Useful Resources:

Use the JINKS method of joinery.
Science textbook units on gears, levers, cams, pneumatics and hydraulics, wheels and axles.
Macaulay, David, *The Way Things Work.*

Integration Connections

Mathematics	Science	Technology
• measurement • scale drawings • calculations • angles • charting and graphing • data gathering	• definition of a variety of forces • types of energy • energy transfer • measuring forces	• open ended problem solving process • materials selection • processing materials • assembly of materials — joining, hinging, fastening • safe use of tools • sketches • working drawings • recyclable materials • strength of assembly • efficiency of operating mechanisms • aesthetics • ergonomics

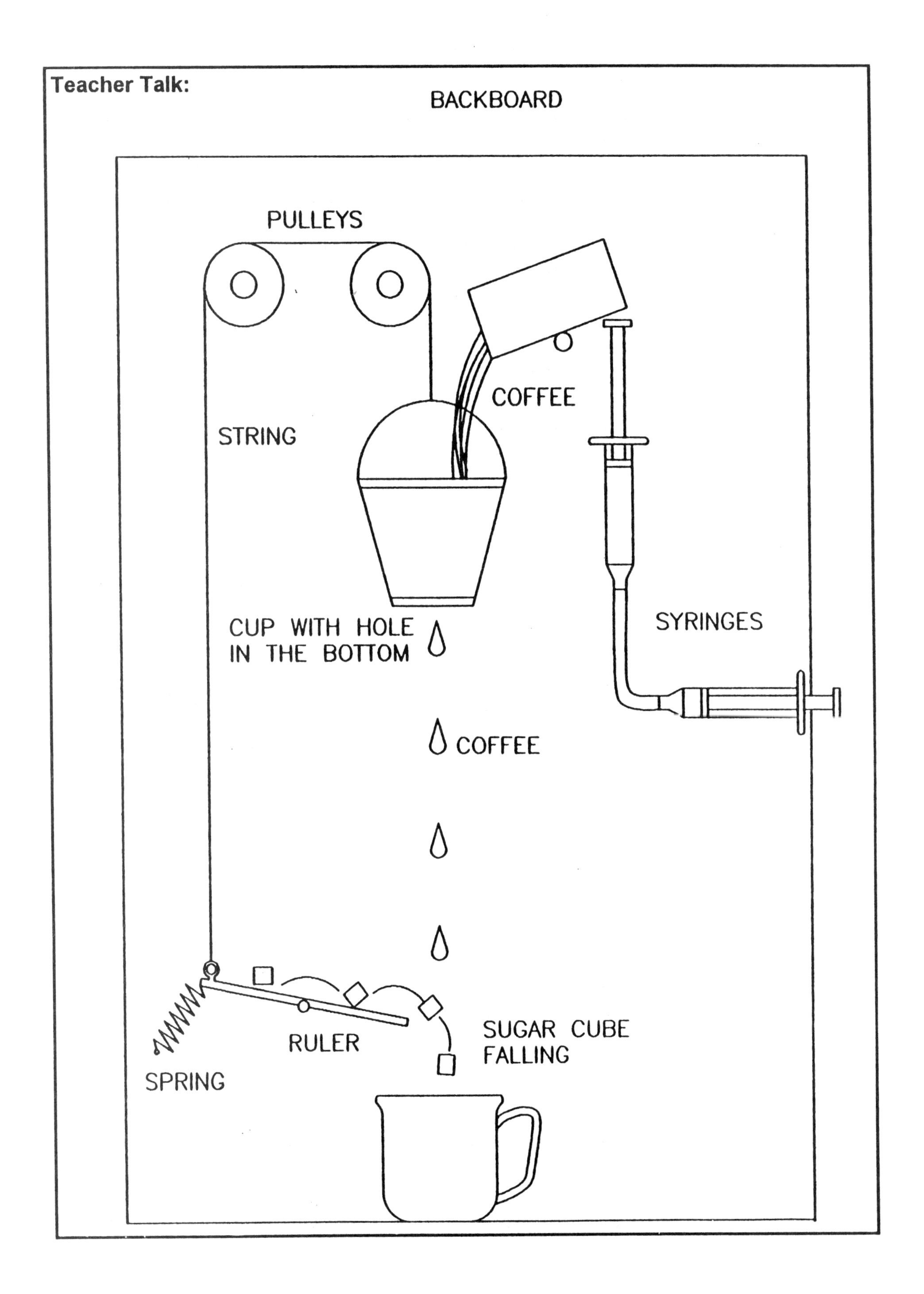
Teacher Talk:
BACKBOARD
PULLEYS
COFFEE
STRING
CUP WITH HOLE
IN THE BOTTOM
SYRINGES
COFFEE
SUGAR CUBE
FALLING
RULER
SPRING

Theme:

STRUCTURES AND MECHANISMS

Title:

"Pump It Up" (Pneumatics)

Integration Connections:

See page 56.

Situation:

You are the chief toy designer for a major toy manufacturer. The president asked your design team for a new line of toys that incorporate an interesting motion controlled by compressed air. The toy should be appealing to young children.

Challenge:

Design and construct a model of a toy that demonstrates motion created by the use of syringes and tubing.

Materials & Equipment:

Could include:
1 cm x 1 cm wood pieces, Bristol board triangles, yellow wood glue (jinks method), dowel, wheels, syringes, tubing, masking tape, scissors, glue brush, miter saw, tri-square, orjinks corner joiner.

Parameters:

Students should work in small collaborative groups.
The toy should not exceed 40 cm in any one direction.
The toy should be safe with no sharp points or edges.
The toy should be durable for use by young children (5-8 year olds).
The toy should be aesthetically pleasing as well as easy to operate.
Timeframe, 5 hours of class time.
Availability of materials and cost factors might be included.

Exploring Ideas:

Will my solution be an animal, person, game, vehicle, or ....?
Will my solution have parts of it move (such as arms on a scarecrow or doors on a vehicle) or will it have steering, or will it propel itself and move across the table?

Choosing & Building the Solution:

Our example: a toy to attempt to trap a balanced ping-pong ball.

Reflections:

Did the toy satisfy all the conditions in the Challenge?
What did we learn by doing this Challenge?
What problems occurred in the process?
Did the group members share responsibilities?
What other resources would have made the Challenge easier to complete?
If you were to do it again, what would you do differently?

Extensions:

How would you market your product?
Create a pamphlet to reflect the qualities of the toy.
What careers could you associate with this Challenge? Investigate them further.
Field trip to ....
Cost out your toy for both materials and labor.
What types of movement can be created by syringes?
How can you increase the force when using syringes?

Some Useful Resources:

Holmes, P.J. & Marshall, D.F., *Move It Further with Air and Water*, Chesterfield.
Mason, Helen, *Great Careers for People Who Like Being Outdoors*.
Czerneda, Julie, *Great Careers for People Who Like to Work with Their Hands*.
Richardson, Peter & Richardson, Bob, *Great Careers for People Interested in How Things Work*.
Lang, Jim, *Great Careers for People Who Want to Be Entrepreneurs*.

Integration Connections

Mathematics	Science	Technology
• linear measurement • cost analysis (extension) • multiplication and division of decimals • calculating volumes of cylinders (extension) • scaled drawings • angles • final product with respect to linear measurement and scaled drawings — is it accurate?	• definition of force • SI unit of force (N-Newton) • measuring forces (using spring scale) • identify and explain the type of movement demonstrated by your toy (linear, rotational, reciprocating)	• open ended problem solving process • materials selection • proccssing materials • assembly of materials — joining, hinging, fastening • safe use of tools • strength of assembly • efficiency of operating mechanisms • aesthetics • ergonomics

Teacher Talk:

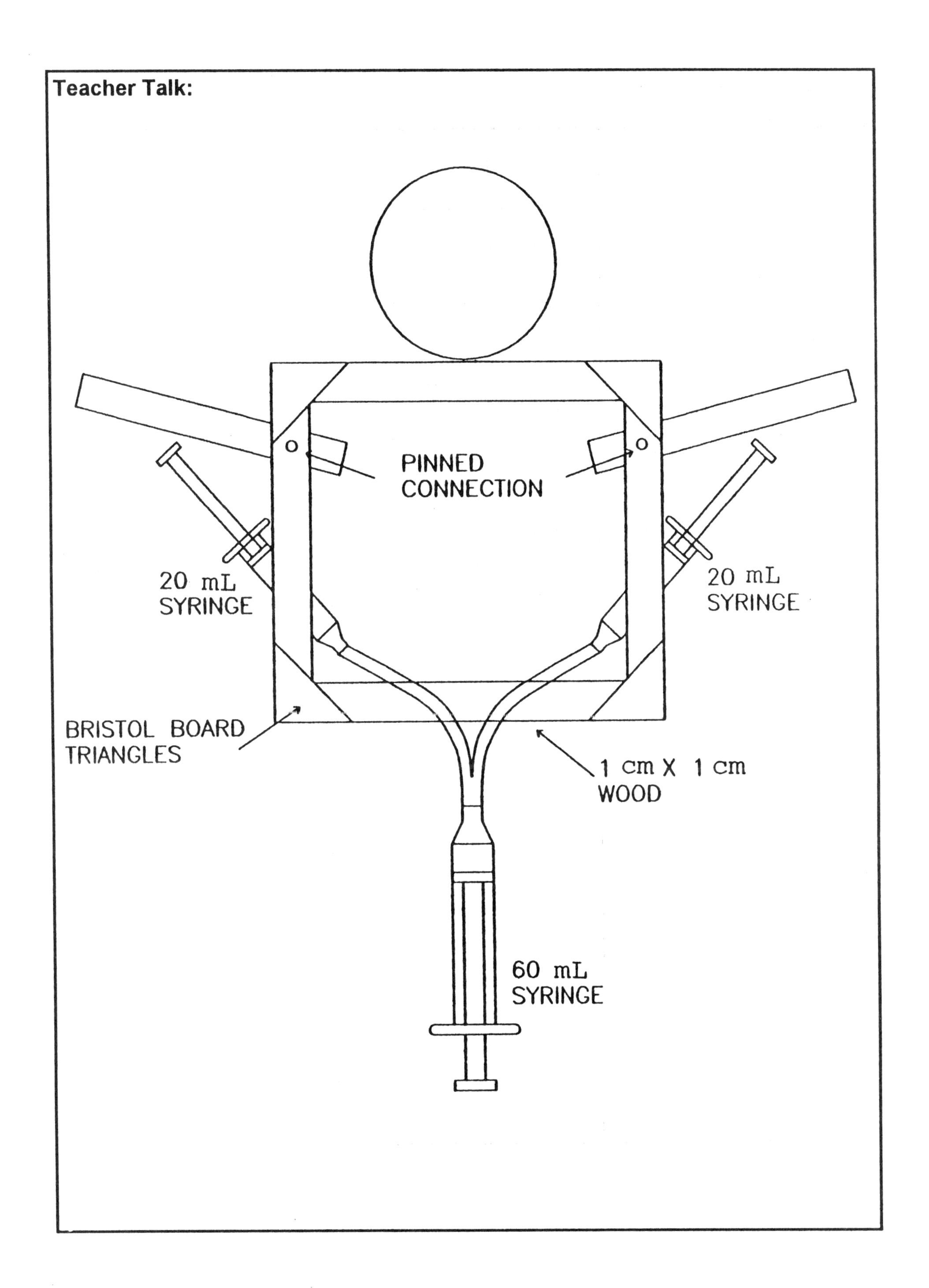

Movement Challenges

Introducing Challenges

Before you begin this section, please consider *Questions to Ask at the Planning Stage* on pages 21-23. Space is left below for your notes on how you might change your strategy next time or perhaps to record in which classes you have used each Challenge, etc.

Teacher Notes

Theme:

MOVEMENT

Title:

"Let The Dog Out!"

Integration Connections: See page 62.

Situation:

You are a designer of BETTER HOMES PRODUCTS and you have been assigned the task of inventing a "dog door". A recent newspaper report suggested a need for an extravagant device suitable for the "rich & famous" market.

Challenge:

Design and construct a model of a door to let a dog outside; the door can be operated from a remote location (e.g., the door will be activated from another room).

Materials & Equipment:

Lego 1030 or 1032 Kits plus pneumatic components may be used to explore ideas.
FINAL SOLUTION may incorporate any of the following: 1 cm x 1 cm wood & gussets, cardboard, thread/rope, gears, pulleys, dowel, nails and/or pins, syringes (various sizes), tubing, valves, electric motors, wire, switches.
TOOLS & KNOW HOW: saws, drills, glue, glue gun, cutting pliers, tape.

Parameters:

Suggest full size dimensions (DOG DOOR 40 cm x 40 cm but suggest working in 1/4 scale (model of door would be 10 cm x 10 cm).
Work in groups of 2 or 3 students only.
The door should be easy to operate and work the first time and every time (reliability is important). Suggested time 5-8 hours.
Attempt to keep the working model as simple as possible and add extensions once the best solution is found (e.g., closing, no access to other animals, etc.).

Exploring Ideas:

Will the door open when the mechanism is activated? Will it operate under repeated trials?
Explore alternatives:
- will the door lift, slide, swing, or rotate?
- will it operate mechanically, with pneumatics, or have electrical components?
- look at different linkages which allows the door to operate smoothly and open wide enough.

Is the door weatherproof so a minimum of heat or cold can escape?
Does the action employ levers, pulleys, pneumatics, and/or electricity?

Choosing & Building the Solution:

Model Various Solutions — create a sketch, drawing, lego model, or mock-up of working parts.
Test different alternatives.
Develop the best solutions and construct a working model.

Reflections:

Does the door operate smoothly and respond as intended?
In what ways could you simplify your solution and still have it perform adequately?
What could you add to your solution so it will work better; what other materials may assist someone with this design problem?
What Mathematics or Science ideas are required as background knowledge for your design?
What advice would you give to another group solving this problem?

Extensions:

How could you restrict entrance to one specific dog only?
Add an automatic device which allows a particular dog to be trained to operate the door but not allow other animal to use it.
Does the door close automatically?
Modify the door to minimize heat loss from the house.
Explore a "revolving door" solution to the problem.
Adapt your door mechanism to be used in a toy "JACK IN THE BOX" or other applications.

Some Useful Resources:

Lego 1030 & 1032 Kits, pneumatic controls and activity cards (see Sample Supplies)
Holmes, P.J. & Marshall, D.F., *Move It Further with Air and Water*, Chesterfield.
Kellett, Joe & Jinks, David, *Design and Make Activities Series*, DJK Technology, 1993.
Corney, Bob & Dale, Norman, *Technology I.D.E.A.S.*

Integration Connections

Mathematics	Science	Technology
• problem solving strategies • ratios • geometric shapes • modeling, sketching, use of scales • scale drawings • alignment and accuracy — measurement • tolerances (percentage error) • how would greater accuracy improve your product? • could the desired movement be accomplished easier (e.g., move fulcrum)? • appropriate calculations of openings	• mechanical advantage and simple machines (levers, pulleys, gears) • measure force required to operate mechanism • compressibility of air	• appropriate application of mechanisms (arms, pulleys, gears) • pneumatics • electric motors — switching and gearing down • appropriate use of tools and materials • economy of constructions • follow a plan — assembly, joinery, mechanical linkages • does it work? • is it reliable? • is it safe for dog and owner? • can you predict where the mechanism may fail? • can it be strengthened?

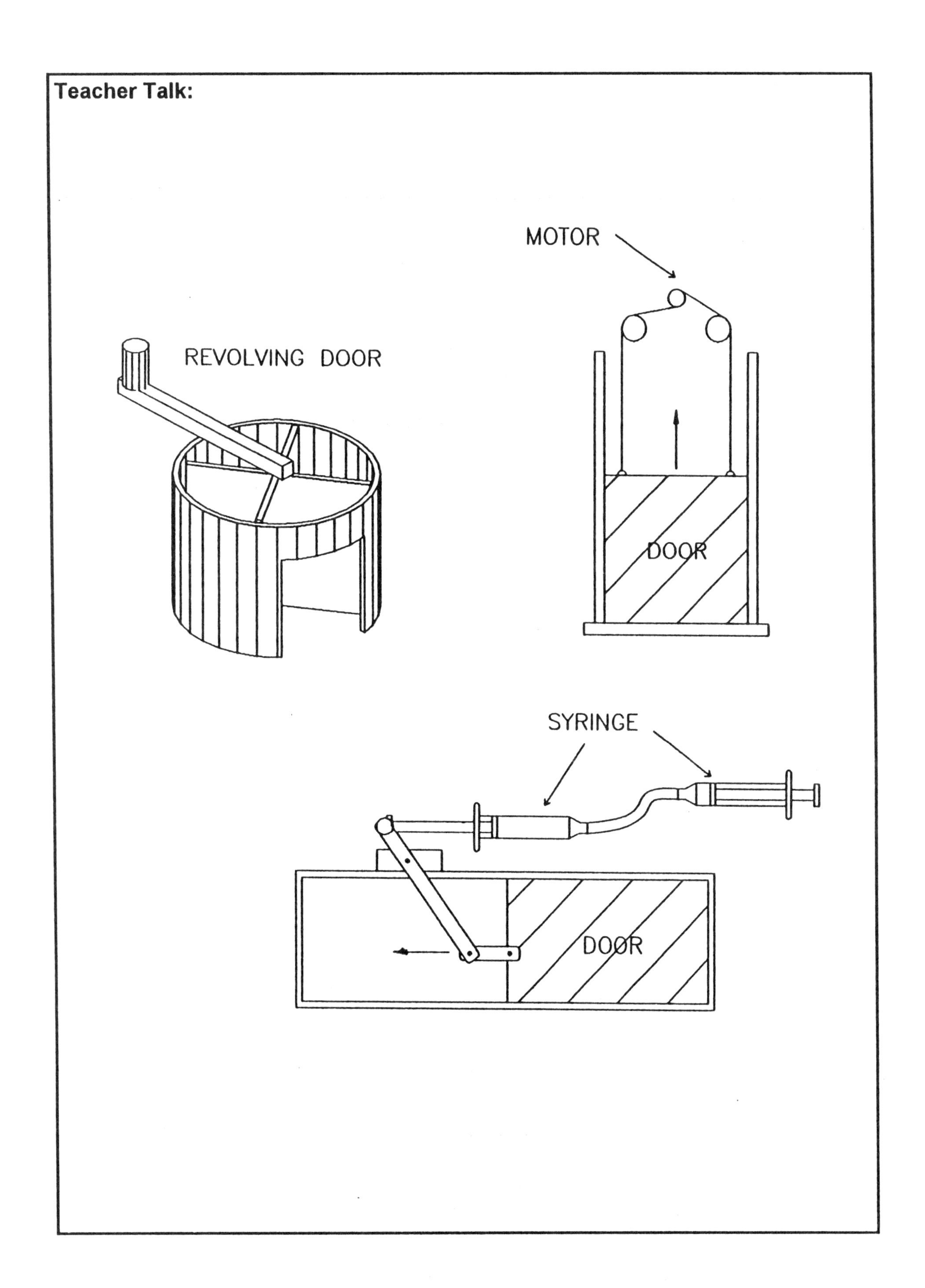
Teacher Talk:
MOTOR
REVOLVING DOOR
DOOR
SYRINGE
DOOR

Theme:

MOVEMENT

Title:

"Star Drop 1: Alien Delivery"

Integration Connections:

See page 66.

Situation:

You are an engineer with an International Space Exploration Agency. You have been asked to design and build a vehicle to drop scientific equipment on a remote earth-like planet.

Challenge:

Design and build a vehicle which will use gravity (only). Your vehicle should rely on wings to make it fall slowly. The best vehicle will fall the slowest from a height of 2 m and land right side up.

Materials & Equipment:

clear plastic 250 mL (8 oz.) drink bottles
paper, and aluminum plates,
polystyrene, acrylic sheets, etc.
Bristol board or other stiff paper
drinking straws
paper clips
possible sticks, balsa wood, etc.
scissors
glue and other adhesives
stop watch or timer
other lightweight recyclable materials

Parameters:

Vehicle sizes may be controlled by limiting the surface areas of wings, the volume may be limited to 60 cm x 30 cm x 30 cm.
Vehicle must not harm the environment, operate safely, and use at least some recycled material.
Suggested time 3-5 hours.
Vehicle starts at a specific height (2 m) and uses no power except gravity.
A tennis ball must be secured to this vehicle without the danger of falling off.

Exploring Ideas:

Will my vehicle explore:
Differences in construction materials and adhesives;
Differences in dimensions, shapes, and number of wings;
Differences in types of chassis;
Vertical drops, angular drops, and combinations of these.

Choosing & Building the Solution:

Look at different construction materials and adhesives, chassis and wings designs, and construction possibilities.
Sketch a model of your vehicle.
Construct, test, and modify until the best solution is found.

Reflections:

Did your vehicle perform as required?
How did you deal with the problems of "Balance in Flight", drag, or air resistance?
What other materials or resources would have assisted you in this Challenge?
What scientific skills did you use to complete this Challenge?
What mathematical skills did you use to complete this Challenge?

Extensions:

The Challenge could be varied so that students may experiment with parachutes.
Vary the payload — the ultimate test would be to use an egg.
Vary the energy source used by the vehicle (e.g., use elastic or spring energy, electricity, etc.).
Explore the applications of your vehicle to other purposes.
Consider the implications of forming a company to market this vehicle.

Some Useful Resources:

Video: "*Gravity*" (Level: grades 7-10), 20 min., color, 1988.
Film: "*Falling Bodies and Projective Motion*" (Level: grades 4-6, grades 7-9, and grades 10-12), 19 min., color, 1984.
Video: "*Dave Hazelwood — Design and Testing of an Aircraft*" (Level: grades 7-9 and grades 10-12), 14 min., color, 1983.

Integration Connections

Mathematics	Science	Technology
• calculating areas of irregularly shaped surfaces • making accurate measurements (areas, volumes, time, etc.) • finding the center of geometrical shapes for attaching wings, etc. • data collection and display (graphs, tables, etc.) • analysis of data patterns and relationships	• hypothesis testing • forces affecting objects in flight — especially gravity • wind resistance or drag • making reasonable interpretations, predictions, and modifications based on relevant observations	• explore geometrical shapes, concrete materials, computers, and other equipment for constructing prototypes • explore appropriate bearing materials (between moveable wings and payload) • use of tools and equipment • working with construction materials and adhesives • timing mechanisms • are there other more suitable resources/ materials?

Teacher Talk:

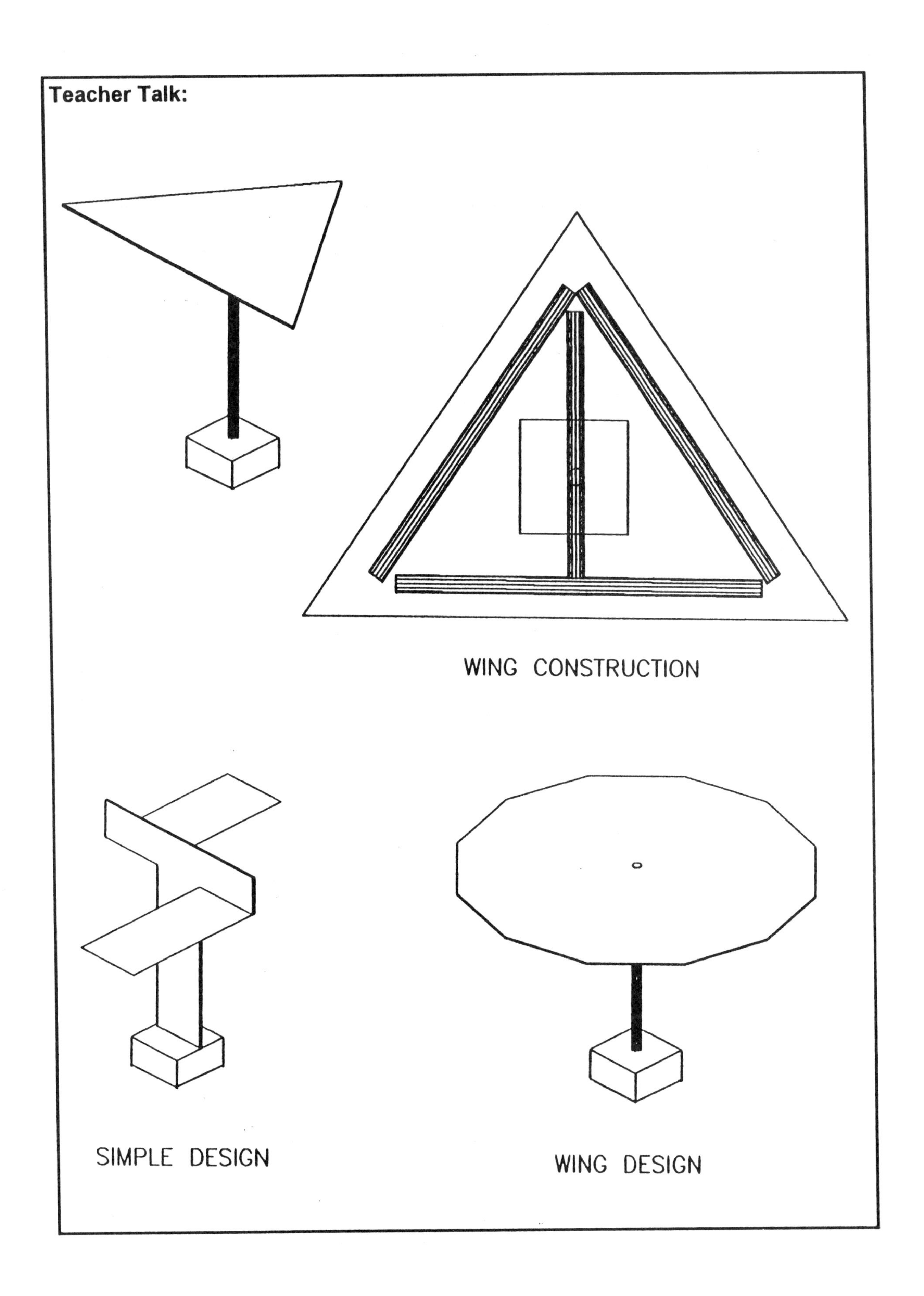

Theme:

MOVEMENT

Title:

"Blowin' in the Wind"

Integration Connections:

See page 70.

Situation:

You are in charge of research and development for a large toy manufacturing company. Your company president, H.I. Lee-Strung, is anxious to develop a new line of musical toys for the highly competitive, pre-school age market.

Challenge:

Design and build the prototype for this new line of musical toys. It must produce a variety of pleasing musical tones using wind energy. Also, it must be safe enough for, and visually appealing to, young children.

Materials & Equipment:

Sound sources:	- wood, plastic, metal or...
Shapes:	- rings, sheets, tubes, pipes, pots, silverware, or...
Other materials:	- wood, plywood, cardboard, dowels, felt, carpet padding, weather-stripping, glue, soft rope, string, picture wire, fishing line, screws, nails
Tools	- small hand tools, saws, drills, glue gun

Parameters:

It should be designed and built by teams of 2-3 students.
The toy should weigh less than 2.275 kg.
It must be wind driven. Suggested time 5-10 hours.
It must produce from 5-12 different musical tones.
It must be sturdy; but, without sharp edges/corners or rough surfaces.
Recyclable materials must be considered for use in constructing this toy.

Exploring Ideas:

What materials should I choose to create sound? How does the sound vary with different sizes and shapes of the material?
How will the wind be produced (natural, fan, paddle, etc.)?
How will the "striker" be activated?
Will the toy sit or hang? Where will the toy be located (inside/outside/floor/wall/ceiling)?
How will the appearance of the toy create a more appealing product?

Choosing & Building the Solution:

Sketch at least two possible designs. Identify the advantages/disadvantages of each design to help make an appropriate choice. Students may allocate the construction tasks based on individual skills and interests.

Reflections:

Did your prototype meet your specifications?
How closely did the musical tones produced approximate a musical scale?
What mathematics or science skills did you use in completing this Challenge?
What other materials or resources would have assisted you in completing this Challenge?
What aspects of your toy will make it competitive in the toy market?

Extensions:

Automate your toy using electrical power.
Remove the restriction on weight.
Designate the toy as "stabile" or "mobile".
Specify that the sounds produced be those of an 8-tone western scale.
Identify ways in which the toy could be made more challenging for the child to interact with.
Survey small children (e.g. nursery or kindergarten class) to identify features that could be used in design.
Research what makes a child's toy "unsafe."

Some Useful Resources:

Sample library resources include:

Roberts, Ronald, *Musical Instruments: Made to Be Played*, Dryad Press, 1972.
Simple Folk Instruments to Make & Play, Simon & Shuster, New York, 1977.

Staff in the music department at your school.
Bartlett, Gillian, *Great Careers for People Interested in the Performing Arts.*
Rising, David, *Great Careers for People Interested in Film, Video, & Photography.*
Czerneda, Julie, *Great Careers for People Who Like to Work with Their Hands.*

Integration Connections

Mathematics	Science	Technology
• measurement • ratios and scale drawings • problem solving • accuracy of measurement and knowledge • different units • total quality of sound • costs of development	• sound and wave patterns • energy • sound vs. material density • observing, measuring • awareness of concepts such as : density, energy, wave patterns • use of a tuning fork	• design process • evaluate different types of materials • various methods of fastening • sequencing of tasks • safe handling of tools • teamwork • ideas for improvement • problems that were encountered and overcome

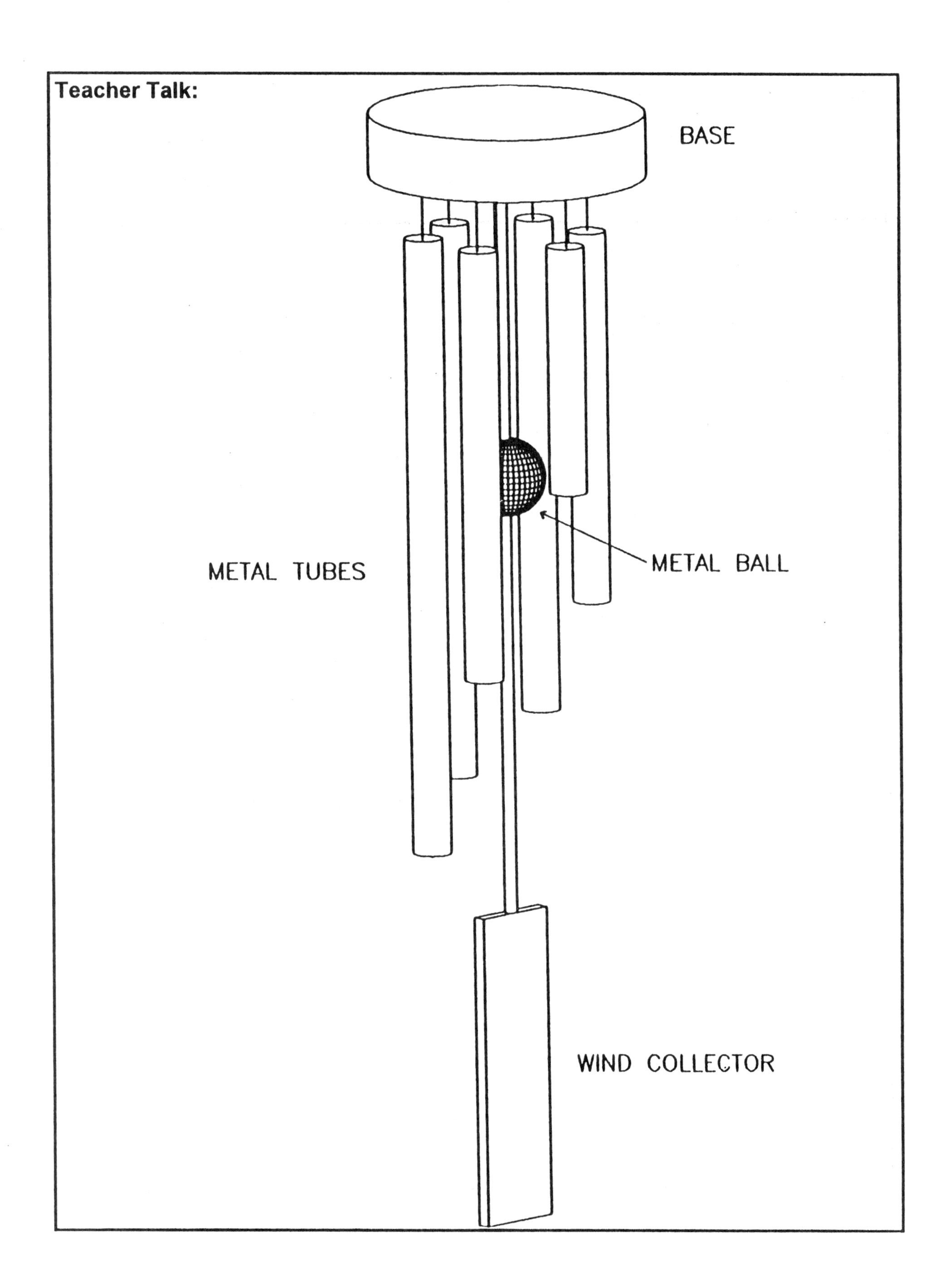
Teacher Talk:
BASE
METAL TUBES
METAL BALL
WIND COLLECTOR

Theme: **MOVEMENT**

Title: **"Go the Distance — Hit the Target"**

Integration Connections: See page 74.

Situation:
As an engineer, you have been assigned the task of designing an UNMANNED vehicle to carry toxic waste down a slope and deposit it at a designated waste recovery site.

Challenge:
Design and construct a working model of a vehicle to carry an unopened can of pop safely down a ramp to a designated spot.

Materials & Equipment:
Ramp Design: (suggested) 1/2 sheet masonite — 50 cm elevation — starting block and target — total distance 8 m
Vehicle Materials: cardboard, 1 cm & 1 cm wood, 20 mm x 100 mm, or 40 mm x 100 mm wood blocks, (1"x 4" or 2"x 4") (suggestions) various balls (tennis, rubber), doweling, skewers, Lego sets, lubrication (wax, oil, grease), rubber bands, cans & bottle tops, numerous recyclable objects, drinking straws
Tools: saws, drills, glue, glue gun, masking tape, electrical tape, duct tape

Parameters: Suggested time 3-6 hours.
The size of the vehicle must not exceed 30 cm x 30 cm x 30 cm and be a maximum of 1 kg without the payload (can of pop).
Student may work individually or groups of 2 or 3 — certain students may be designated as "set-up" and "testing" persons.
The vehicle must operate safely and use at lease some recycled materials.
The vehicle must NOT harm the environment.
A can of pop must be secured on the vehicle without danger of spillage.
The vehicle will start from a stationary position at the top of the ramp and descend using NO power except gravity.
The vehicle MUST come to rest (STOP) at a pre-determined spot (target).

Exploring Ideas:
Examine a variety of different wheels and axle assemblies.
How will the framework and chassis be constructed?
Is lubrication and bearing surface a design consideration?
Is the shape, size, and mass employed the best alternatives?
How can we control the vehicle to STOP at a designated spot?
Will it operate consistently over repeated trials?

Choosing & Building the Solution:

Examine and experiment with different materials, wheel designs, and construction possibilities. Sketch one of your alternatives, construct it, test, and modify the model until the best solution is found.

Reflections:

Did the vehicle perform as expected?
What mathematical skills did you use to complete this Challenge?
List 2 of the most important problems that you encountered. How did you overcome these problems?
What other materials or resources would have assisted you in this Challenge?
List any suggestions you would give to help other vehicle designers.

Extensions:

Vary the ramp — slope, surface texture, height, distance, and size of target, curve the path to the target.
Vary the outcome (target)
- design a measuring device tu measure speed
- have vehicle perform a task at the target (e.g., unload, break a balloon or ?)

Explore other applications for the vehicle and market your product.
List the career areas in which you can apply the skills you have learned.
Cost out your vehicle for materials and labor? How can you economize?

Some Useful Resources:

Found materials: students and teachers get involved in searching out all possible sources
Lego 1030 & 1032 Kits, pneumatic controls and activity cards (see Sample Supplies).
Corney, Bob & Dale, Norman, *Technology I.D.E.A.S.*
Richardson, Peter & Richardson, Bob, *Great Careers for People Interested in Math & Computers*.
Grant, Lesley, *Great Careers for People Concerned About the Environment*.
Lang, Jim, *Great Careers for People Who Want to Be Entrepreneurs*.

Integration Connections

<table>
<tr><th>Mathematics</th><th>Science</th><th>Technology</th></tr>
<tr>
<td>
<ul>
<li>problem solving</li>
<li>linear measurement (unit)</li>
<li>geometric shapes (2 & 3D)</li>
<li>ratios</li>
<li>recording and analyzing data — scale drawings</li>
<li>record keeping and sketching (possible computer applications)</li>
<li>accuracy of measurement</li>
<li>estimation and understanding units of measurement</li>
<li>costing — decimal arithmetic</li>
<li>are the measurements accurate?</li>
<li>would improved measuring techniques improve performance?</li>
</ul>
</td>
<td>
<ul>
<li>force and gravity/momentum</li>
<li>friction</li>
<li>measure and record variables, e.g., mass</li>
<li>identify the forces involved with the vehicle movement</li>
</ul>
</td>
<td>
<ul>
<li>types of wheels</li>
<li>material selection</li>
<li>bearings and lubrication</li>
<li>sketching</li>
<li>follow a plan</li>
<li>demonstrate appropriate use of equipment</li>
<li>assembly, fasteners, joinery</li>
<li>economical use of materials</li>
<li>did your product achieve the desired goal?</li>
<li>how could the design be improved?</li>
</ul>
</td>
</tr>
</table>

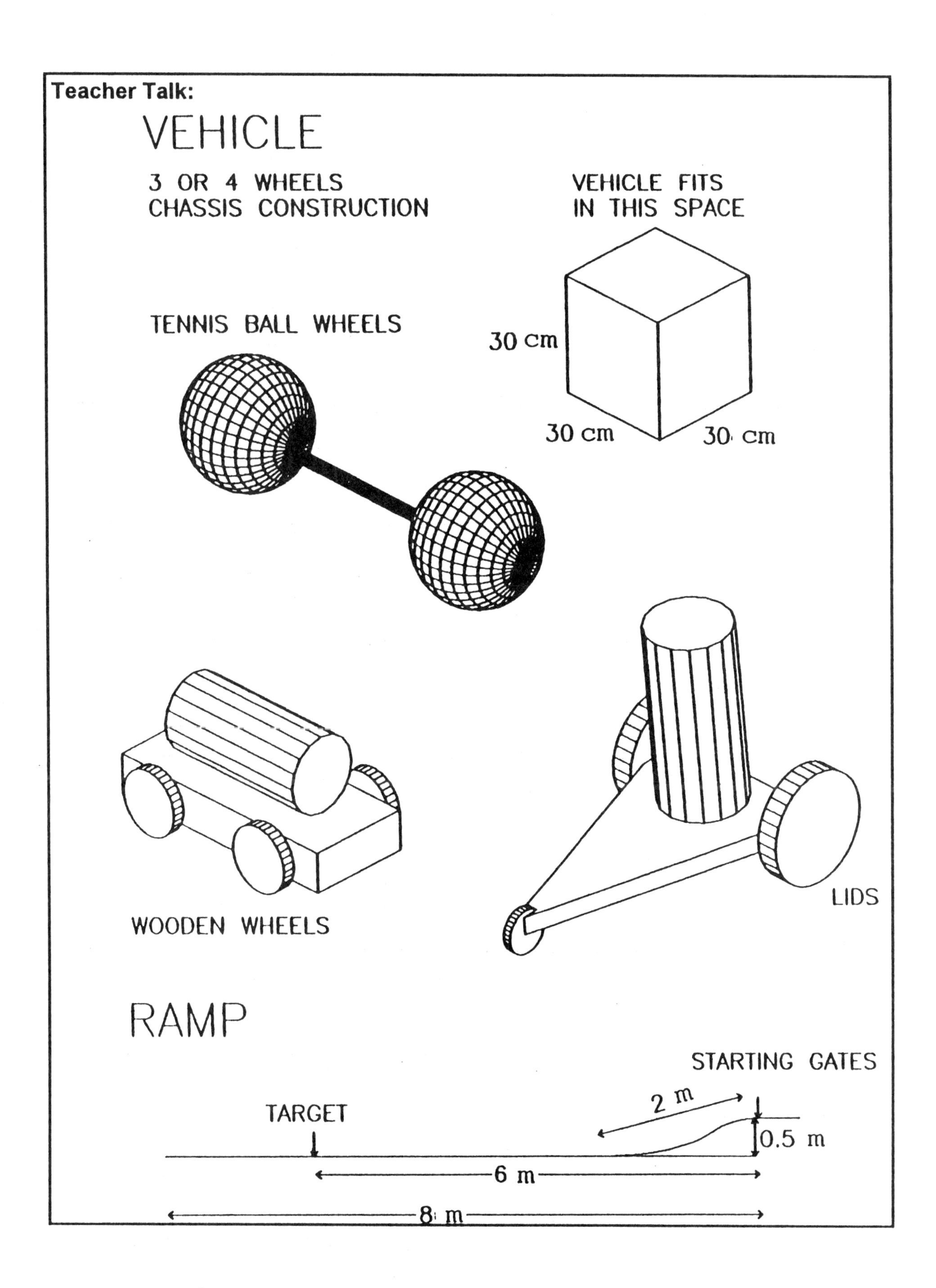
Teacher Talk:
VEHICLE
3 OR 4 WHEELS
CHASSIS CONSTRUCTION
VEHICLE FITS
IN THIS SPACE
TENNIS BALL WHEELS
30 cm
30 cm
30 cm
WOODEN WHEELS
LIDS
RAMP
STARTING GATES
TARGET
2 m
0.5 m
6 m
8 m

Theme:

MOVEMENT

Title:

"The Futuremobile"

Integration Connections:

See page 78.

Situation:

It is the beginning of the 21st century. Fossil fuels are rare and very expensive. As an inventor, you have experimented with several ways to propel small vehicles. You have considered electricity, wind power, solar power, and compressed gas as possible sources of energy for vehicles.

Challenge:

Select one source of energy.
Design and build a working model of a vehicle using this energy source.
Demonstrate the advantages of your design to potential customers.

Materials & Equipment:

Suggested materials: wood, cardboard, sheet acrylic, sheet aluminum, doweling, 3 mm steel, or aluminum rod. Wheels may be purchased or hand-made. Give consideration to recyclable materials such as peanut butter jar lids, soft drink bottles, tin cans, etc.
Decorating/painting with a futuristic design is recommended.
Tools: small hand tools, saws, drills, glue gun, lathes, or mill, depending on level of student skill and equipment available.

Parameters:

Must be designed and built by a group of three or four students in 5-10 hours.
The model must be self powered over a level surface for at least 10 m.
The vehicle must measure no more than 40 cm in any direction.
A guide wire or other directional control may be used.

Exploring Ideas:

Which energy source will we use? Is there an environmental consideration?
How will energy be converted to forward motion?
How heavy/light should our model be?
What materials will best meet the strength and size requirements?
What advantages are offered by our design?
What limitations exist in our tools and equipment?

Choosing & Building the Solution:

Several scenarios should be explored with sketches and notes for each. Discussion of the pros/cons of each suggested design will enable the group to make a more informed choice. Some "mock-ups"/experimentation may be necessary. As a team, students may assign construction tasks based on individual skills/interest.

Reflections:

How well did the model perform? What improvements could be made?
What was the most difficult problem in selecting the source of energy, the design, and the construction?
What advice could you give to another group working on this Challenge?
Which energy source do you think will be important in the 21st century?
What additional materials or tools would result in a better model?
What futuristic ideas could not be included in your design because of limitations in tools, equipment, or material?

Extensions:

Increase the travel distance required.
Eliminate the "control wire" used for tracking/steering.
Require that the vehicle travel a circular path.
Specify a load that must be carried.
Produce a flyer, advertisement, or video to promote your design.

Some Useful Resources:

Source of CO_2 Cartridges: Large hardware stores
(see Sample Suppliers)

Integration Connections

Mathematics	Science	Technology
• ratios and equations • speed and time • scale • estimation • linear measurement • area • measurement accuracy • tolerances • fractions • graphing results • costing/cost efficiency	• sources of energy • energy conversion • given that a solution to the Challenge is a scaled down prototype, discuss the validity of predictions • explain sources of energy • observation • reasonable interpretation • environmental impact	• evaluation of materials • mechanisms • sketching • the design process • process planning • working safely • assembly methods • using tools and equipment • development of group skills • identification of problems • ideas for improvement

Teacher Talk:

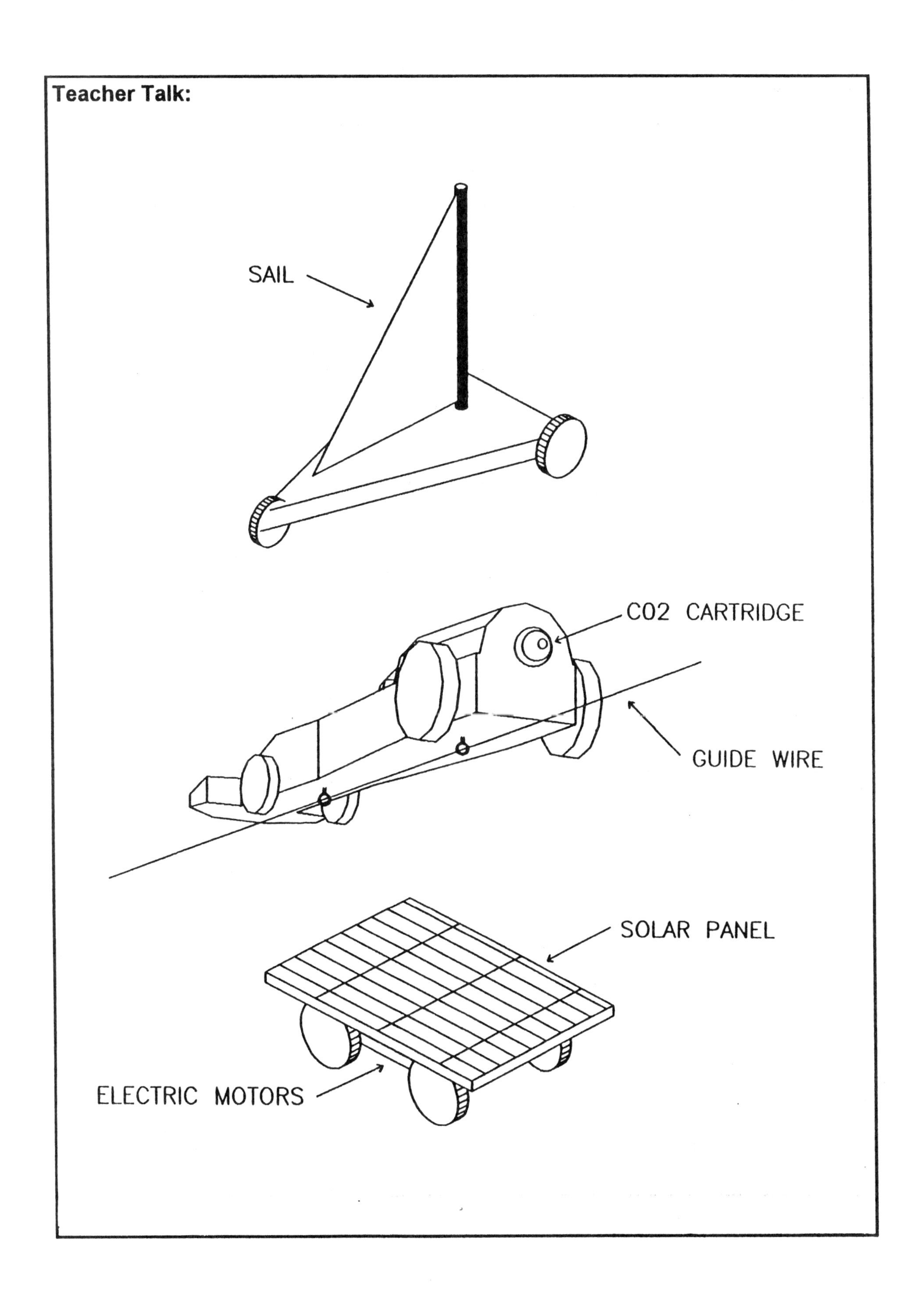

Theme:	MOVEMENT

Title:	"A Risky Move"

Integration Connections:	See page 82.

Situation:
Your business handles a large quantity of chemicals. At one stage in the handling of these chemicals, a barrel of dangerous, flammable liquid must be moved from a truck to a storage rack. The barrel must not be touched by human hands.

Challenge:
Design and build a working model of a hazardous material handling system that will move a barrel from a truck to a storage rack.
Design and build a suitable storage facility.

Materials & Equipment:
The Storage Rack: shelf materials may include stiff cardboard or wood, sheet metal, plastic, etc.
The Mechanism: materials may include some or all of 1 cm x 1 cm wood, cardboard, aluminum sheet, or angle, syringes and tubing, hydraulic and/or pneumatic cylinders, light cord or cable, pulleys and gears, and an assortment of fastening materials
Tools: small hand tools, saws, drills, glue gun, lathes, or mill, depending on level of student skill and equipment available

Parameters:
Must be designed and built by a team of two or three students in 8-10 hours.
The "barrel" of toxic material is a 284 mL (8 oz) tin can.
At least two hydraulic or pneumatic components must be used.
The mechanism must be able to reach 30 cm, grasp, lift to a height of 5 cm, move, and deposit the "barrel" on the storage rack, then return for another "barrel".
At least one innovation for environmental protection must be included in the design of the mechanism or storage facility.

Exploring Ideas:
What will function as a base that will allow the mechanism to operate without tipping?
How will the "barrel" be grasped (forks, jaws, magnet, etc.)?
What mechanisms can provide 5 cm of lift?
How will the mechanism move the "barrel" from its initial position to the storage area?
How will the "barrel" be deposited in storage?
What type of storage facility will provide safe storage and easy access?

Choosing & Building the Solution:

Consideration should be given to at least three potential solutions using a variety of materials. Discussion of and listing the pros/cons of each possible solution will enable the group to make a more informed choice. As a group, identify the steps and processes involved in construction of the model, then allocate tasks based on individual skills and interest. One member of the group may act as co-ordinator.

Reflections:

How well did the model perform? What improvements could be made?
What was the most difficult problem in 1) design, 2) construction?
What advice could you give to another group faced with the same Challenge?
What might be other environmental concerns?
What safety concerns might exist in a full-size operation?

Extensions:

Change weight, size, distance, etc.
Increase the weight of the "barrel" or the distance it has to be moved.
Introduce a requirement for positioning accuracy.
Design and build hydraulic/pneumatic cylinders. This could involve the use of copper pipe, aluminum rod, "O" rings, soldering, etc.
Alter the goal, e.g., use two or three shelves and deliver the "barrel" to any designated shelf.

Some Useful Resources:

Holmes, P.J. & Marshall, D.F., *Move It Further with Air and Water*, Chesterfield.
Kellett, Joe & Jinks, David, *Design and Make Activities Series*, DJK Technology, 1993.

Integration Connections

Mathematics	Science	Technology
• transformational geometry, e.g., translation, rotation • simple equation applications • measurement: angular, linear • surface area, volume • tolerances • slope • performance accuracy	• pressure/volume relationships • levers • simple machines • force, work, energy • does a scale model give any evidence in assessing an environmental impact? • does a scale model allow us to determine if the system is efficient and effective?	• evaluation of materials • hydraulics/ pneumatics • mechanisms • the design process • process planning • drawing, documentation • working safely • assembly methods • using tools and equipment • develop group skills • identification of problems • ideas for improvement • assess work-place safety of the system

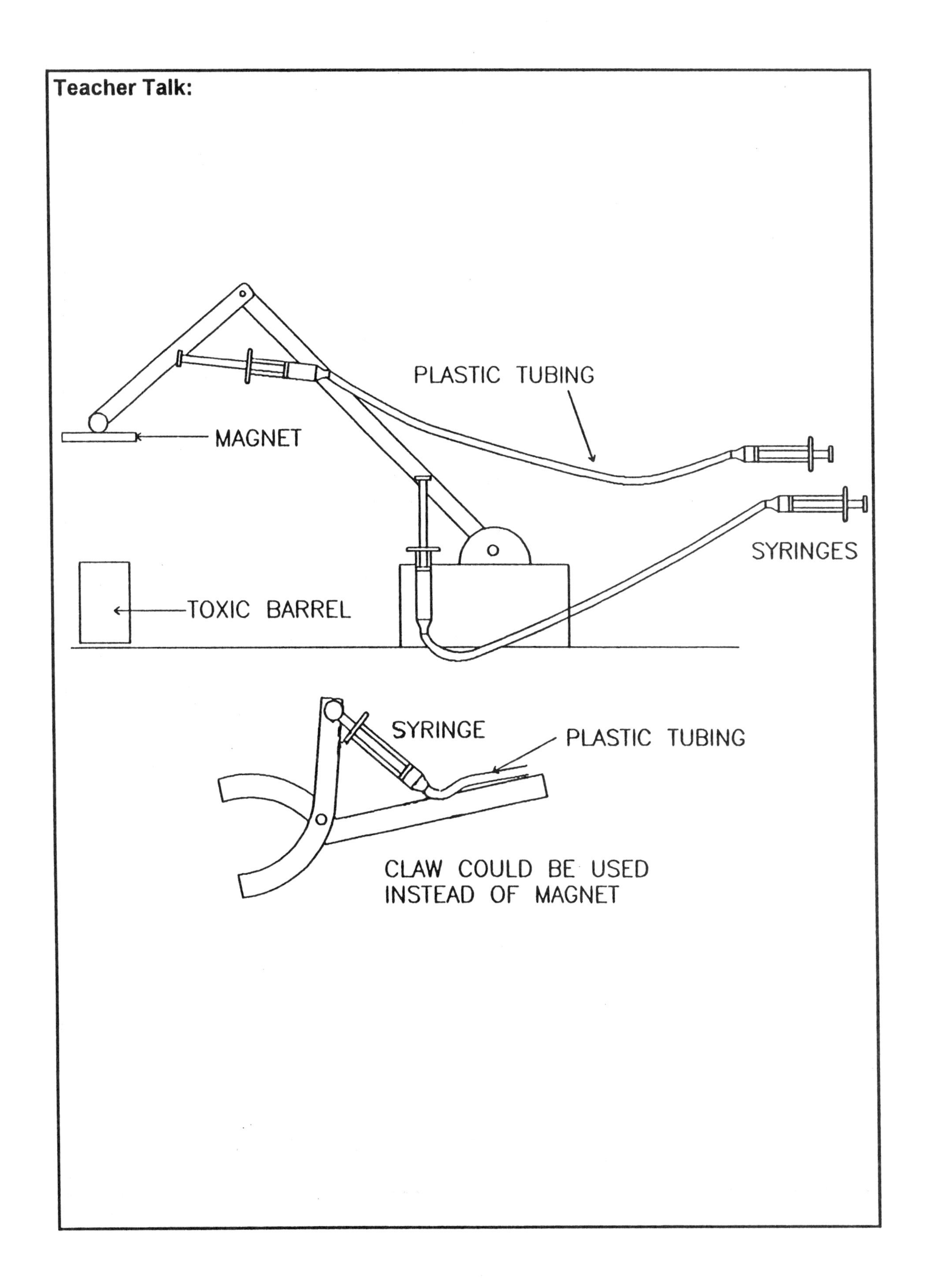
Teacher Talk:
PLASTIC TUBING
MAGNET
SYRINGES
TOXIC BARREL
SYRINGE
PLASTIC TUBING
CLAW COULD BE USED
INSTEAD OF MAGNET

Force Challenges

Introducing Challenges

Before you begin this section, please consider *Questions to Ask at the Planning Stage* on pages 21-23. Space is left below for your notes on how you might change your strategy next time or perhaps to record in which classes you have used each Challenge, etc.

Teacher Notes

Theme:

FORCE

Title:

"Welcome to the Fast Lane"

Integration Connections:

See page 88.

Situation:

As a member of an international racing design team, your sponsor requests that your team design a car to win the next race or sponsorship will be withdrawn.

Challenge:

Design and construct a model from the materials provided, to travel as fast as possible, following the TSA 500 Dragster specifications.

Materials & Equipment:

wood block (45 mm x 70 mm x 300 mm.)
2 metal axles
2 drinking straw bearings
4 wheels
4 washers
paint

TSA Dragster specs.
drill press/hand drill
scroll saw
files/sandpaper
paint brushes
CO_2 cartridge

Parameters:

Students will work in design teams of no more than four. However, individuals will construct their own car for selection by the group to compete in the race.
All teams receive and use the same material.
Time frame, approximately 15 hours.

Exploring Ideas:

Working within groups or with a partner:
Research the factors that affect speed;
Generate thumbnail sketches (at least three);
Research suitable finishes to reduce friction and drag;
Research how a fair test can be set up to race the dragsters;
Use CD ROM and other publications in their exploration.

Choosing & Building the Solution:

From the best sketch:
Complete a top and side view scaled drawing;
Construct the dragster;
Revise drawings as required;
Select from within the group, the best car for the race.

Reflections:

Did the dragster satisfy all the conditions in the Challenge?
What did we learn from the process?
What are the connections to Mathematics and Science?
What safety regulations were observed?
What problem solving techniques were used?
What human processes were involved?

Extensions:

Write a technical instruction pamphlet or report on how to build the dragster.
Can we use an alternate type of propulsion, e.g., solid versus liquid fuels?
Reduce parameters to allow for more creativity.
What careers can we associate with this Challenge?
Investigate cost of construction.

Some Useful Resources:

Hacker & Borden, *Living with Technology*.
Metric Dragster Kit (see Sample Suppliers)
Car Builder Software (see Sample Suppliers)
Source of CO_2 Cartridges: local hardware store or scientific supply house.
Any high school physics textbook
Lang, Jim, *Great Careers for People Who Want to Be Entrepreneurs*.
Richardson, Peter & Richardson, Bob, *Great Careers for People Interested in How Things Work*.
Vincent, Victoria, *Great Careers for People Interested in the Past*.

Integration Connections

Mathematics	Science	Technology
• geometry — locating center of a circle • symmetry • calculate number of wheel rotations • $\text{velocity} = \frac{\text{distance}}{\text{time}}$ • graph (distance, time) • scale drawing • perspective drawings • calculating unit costs	• definition & identification of forces • potential/kinetic energy • energy conversion (conversion of energy) • propulsion, thrust, traction, friction, gravity, drag, speed, distance, acceleration • aerodynamic shapes • CO_2 under pressure • temperature change/gas • factors affecting friction (mass, surface type, surface area, rollers) • measurement of force • inertia • volume of wood used compared to volume of wood given (immersion in overflow cylinder)	• research (library & books) • design sketches • examine problem solving processes • prototype building • scaled working drawings • materials selection • safe use of tools and equipment • cutting curves • filing, shaping, sanding • finishing methods and materials • assembly of parts • testing efficiency of rotating parts • aerodynamics test • race day performance • group dynamics • report writing • aesthetics

Teacher Talk:

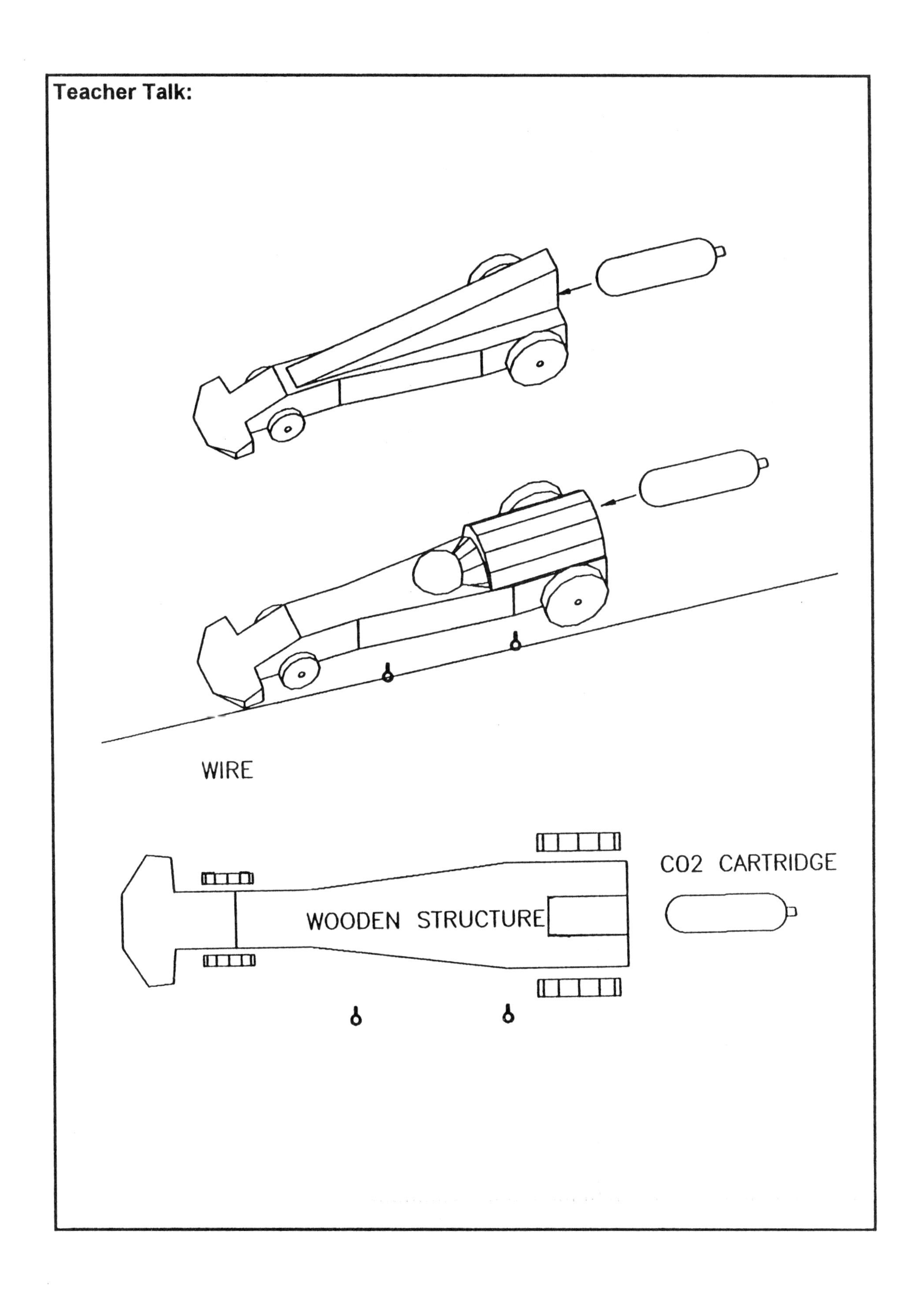

Theme:

FORCE

Title:

"Get A Grip"

Integration Connections:

See page 92.

Situation:

As a member of a medical equipment design team, the hospital administrator requests that your team design a "grabber" to assist individuals with reaching and lifting.

Challenge:

Design and construct a model which will use pressurized air as the means of control to lift an irregular shaped object using the materials provided.

Materials & Equipment:

2 syringes (30 cc x 15 cc)
1 m 6 mm plastic tubing
1 m 1/2 in. dowel or substitute
(PVC tubing, measuring stick)
masking tape
2 cm x 1 cm wood for jaws (500 mm)
2 wood screws
glue gun
drill and bits
sandpaper squares
Bristol board (for triangle supports)

Parameters:

Not to exceed 1 m in length
Must be able to pick up an irregular-shaped object from the ground using pneumatically controlled jaws.
Must use the given materials.
Working with a partner, to determine a solution.

Exploring Ideas:

What is pneumatics/hydraulics?
Investigate uses of air pressure (e.g., air brakes, automatic/sliding doors, dentist/barber chair, garage hoist, fork lift).
Research units of measuring pressure.
Explore the field of ergonomics.
Research other existing devices that are for a similar purpose (e.g., medical supplier).

Choosing & Building the Solution:

Make sketches of possible solutions;
Experiment with alternatives and choose the best solution;
Construct the grabber;
Revise and test when required.

Reflections:

Did your grabber satisfy all the conditions in the Challenge?
Did your grabber perform as expected — if not, can you explain why not?
How well were the responsibilities shared within the partnership?
What other uses can you think of for your grabber?
How could you improve the "grip" — ability of the grabber?
If you were to do it again, what would you do differently?

Extensions:

Design an ad campaign to "sell" your product (other uses for the grabber).
Compare efficiency (practicality) of hydraulic vs. pneumatic (change medium from air to water).
Vary angle of jaws for better grips.
Build a robotic arm using this principle.
Weight-lifting contest?
"Pick up an egg" contest?
Design and build a device that can be single hand operated.
Have a variety of different shapes — whose grabber can lift them all?

Some Useful Resources:

Holmes, P.J. & Marshall, D.F., *Move It Further with Air and Water*, Chesterfield.
Macaulay, David, *The Way Things Work.*
Homecare medical equipment store for ideas re. "grabbers" for disabled
Any high school physics textbook

Integration Connections

Mathematics	Science	Technology
• cost analysis • linear measurement • fractions • scaled drawings • angles • calculating volumes of cylinders (syringes) • measure final product with respect to linear measurement and scaled drawings — is it accurate?	• compression force • liquid vs. air under pressure • simple machines — changing direction of force applied • devise method of testing strength of grabber (e.g., measure depth of indentation in a ball of clay created by grabber claws) • comparing grabbers — is it a fair test? (control variables)	• research (library) • design sketches • problem solving • safe use of hand tools • materials selection • marking and measuring • cutting of pieces • drilling of holes • fitting of syringes • assembly of parts • use of adhesives • ergonomics • test it • vary amount of air in each syringe • increase weight of object being lifted • efficiency of air vs. water/fluids in syringe • group dynamics — what problems were encountered working together? • write a report with illustrations

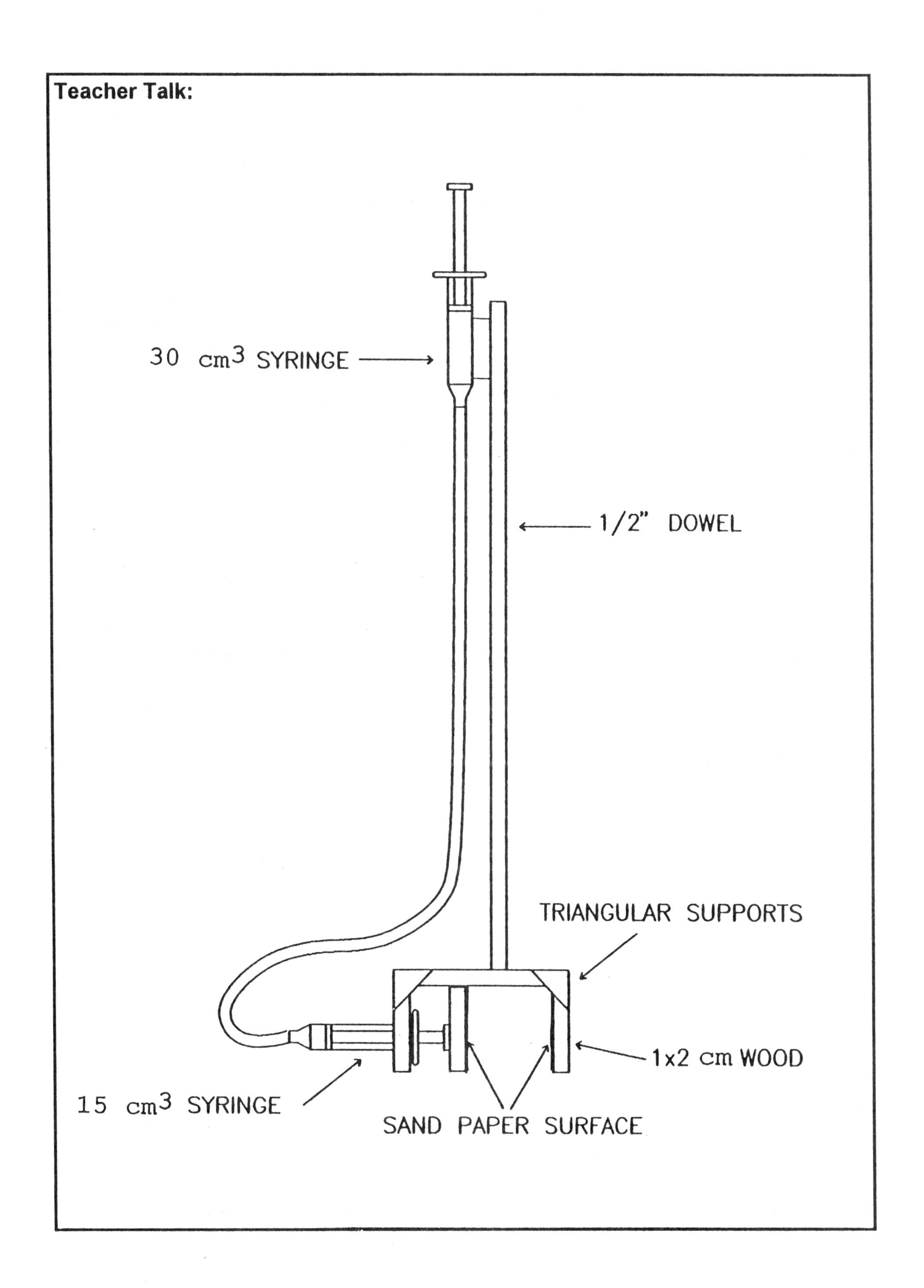
Teacher Talk:
30 cm3 SYRINGE
1/2" DOWEL
TRIANGULAR SUPPORTS
1x2 cm WOOD
15 cm3 SYRINGE
SAND PAPER SURFACE

Theme:

FORCE

Title:

"Take Off!"

Integration Connections:

See page 96.

Situation:

You are stranded on the uninhabited side of a tourist island but cannot escape due to a thick and dangerous jungle between you and civilization. You need to let someone know you're there, but all you can find around you is an old tire, air pump, pop bottle, and scraps of wood and paper litter washed up on the beach.

Challenge:

Design and construct a plastic pop bottle rocket that can be propelled the greatest distance/height.

Materials & Equipment:

2 L (64 oz.) plastic pop bottle(s)
tire valve (e.g., TR-425 by Dill)
air pump with long hose
launcher (see picture for example)
various items for fins/cones, etc.
adhesive materials (duct tape, glue, etc.)
water (inside bottle)

Parameters:

Rocket and all attached parts must fit into launcher properly.
No more than 2 or 3 bottles used.
Groups of 3 or 4 students working together to build rocket.
Use materials listed.
Time frame — building and design — approximately 3 hours.

Exploring Ideas:

Characteristics/features of rockets — why so designed (functions)?
Do we need water in the bottle — if so, how much is best?
What's involved in flight — lift, drag, propulsion, aerodynamics, wind resistance?
How does a valve work?
How do we get so much power just from air (air pressure)?
How will the angle of the launch affect the distance traveled?
Is the launch both fair and safe?
What materials will be best for constructing, considering mass, flexibility, ease of adhesion?

Choosing & Building the Solution:

The group will: brainstorm ideas, produce rough sketches, negotiate and determine the best solution, create working drawings, and construct a rocket.

Reflections:

Did we follow instructions given/make use of advice suggested?
Was our rocket aerodynamically efficient? What improvement could be made?
Did the rocket perform as expected? If not, what might explain the unusual flight/landing/launch?
Watching all the rockets launched, was there one feature common to all the "winning" rockets?
Was our construction sound, i.e., did it survive the launch and landing?
If you were to do it again, what would you do differently?
How would you rate your group's success — what was your role in this?

Extensions:

Can you design a better launcher?
Map the landings of all the rockets — any pattern?
Change a variable while controlling all others to determine the "ultimate" rocket (e.g. water/air ratio, water temperature, nose weight, stabilizer location, cowling length).
Hold a rocket "beauty contest" — creative design and decoration.
Put wheels on the rocket and launch along the ground.
Careers — manufacturing/technology behind the plastic (incredibly strong).

Some Useful Resources:

"Strato Blaster" (see Sample Suppliers)
"Hydro Launch" Hydrokinetic Propulsion System (see Sample Suppliers)
Large hardware or automotive supply store — tire valves (TR-425 by Dill)
Macaulay, David, *The Way Things Work.*
Any high school physics textbook

Integration Connections

Mathematics	Science	Technology
• ratio • measuring angles using protractor • symmetry of design • scale drawings • metric conversion • similar triangles • dividing a circle into equal parts • graphing relationship between angle/distance • calculate averages (flight times, distance, height, etc.) — use of clinometer • volumes, surface area • shapes • extension = formula using tangents	Experiment to determine: • best ratio of air/water and affect/non-affect of performance • variables such as length, nose weight stabilizers (size & location) • properties of matter — adhesives/water • flight (lift, drag) • potential/kinetic energy • aerodynamics/friction of wind resistance • energy conversion/ conservation • consider energy inefficiency due to mass (use of scales) • air pressure • valves • increased force due to cowling around valve to direct water flow kickplate for water to "push off" of • accurate observation/timing of rocket flight (younger grades) — use stop watch & pace measurement	• design sketches showing different perspectives • labeled diagrams including measurements • use of cowling to direct water • materials selection • selection of adhesive on plastic with exposure to water • location of stabilizers/wings, etc. • safe use of shears cutting plastic • stable insertion of valve for launching • attaching rocket features to withstand fit into launch pad and the actual launch/landing • launching of rockets • aerodynamic shape (pointed nose, etc.) • stabilizers — size, number, location, shape • weight to length ratio — nose must be heavier • materials used • condition of rocket after flight

Teacher Talk:

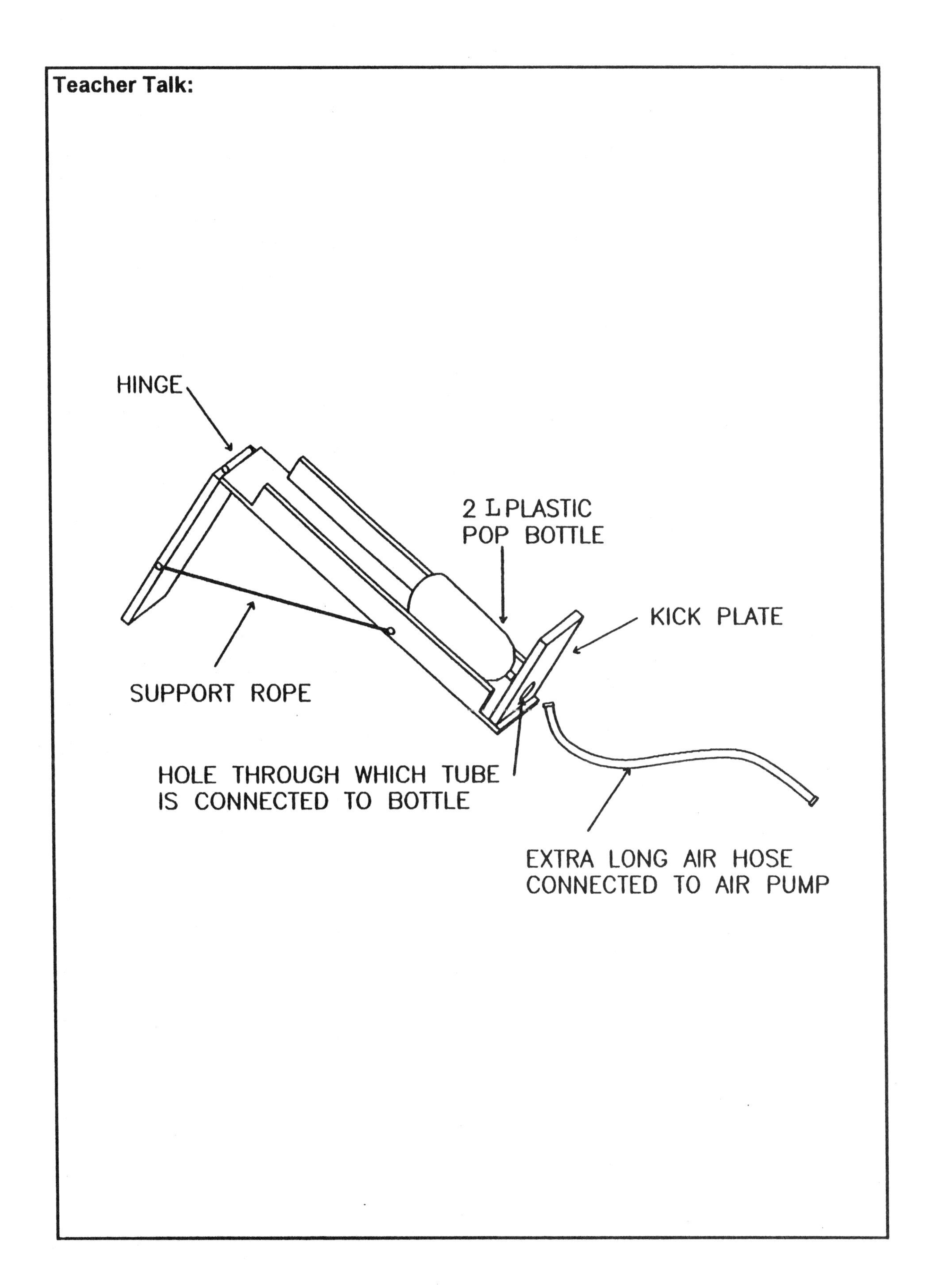

Theme:

FORCE

Title:

"The Catapult — Death by Marshmallow!"

Integration Connections:

See page 100.

Situation:

A toy manufacturer has a need to develop a toy that will propel an object accurately at a target. The device must be safe and have entertaining appeal to kids.

Challenge:

Design and build a device that will launch a marshmallow accurately at a given target, with the use of elastic force.

Materials & Equipment:

Wood — basswood or willow 0.5 x 1 x 1 m lengths; 1/4" dowel; spools
Hand Tools — miter saw or miter box and hack saw; hot glue gun and glue sticks or carpenters glue, measuring tape; hand drill or drill press, hammer, coping saw clamps; 1" nails, corner joiner
Drawing equipment — architect scale, set squares, T square for drafting
Prototype Building — Lego 1030 Kit
Other — elastic bands, large marshmallows

Parameters:

The catapult should:
Launch a large marshmallow accurately at a distance of 2 m;
Be constructed of wood with no single piece in excess of 25 cm;
Have a maximum of 2 nails and 2 elastic bands;
Be of durable construction;
Be constructed in 5 hours of class time;
Be a collaborative effort of 3 students.

Exploring Ideas:

In a group of no larger than 3, assess the situation and Challenge. Brainstorm the design of a catapult device examining through research and application:
Historic catapult designs (mobile & stationary catapults);
Medieval construction technique;
The properties of the given materials (elastics, wood pieces);
The practical, safe, and efficient use of given tools and materials.

Choosing & Building the Solution:

The group will brainstorm sketch ideas, negotiate the best solution, model with a prototype, revise the selected design, create working drawings, construct the final design, prepare a presentation package incorporating all the sketches, drawings, and construction details from above.

Reflections:

Applications to sports, mousetraps?
Were all the safety considerations recognized?
Was the catapult efficient? Did it launch successfully without damage to the device?
What improvements could be made to the design or construction?
What made a successful catapult?
What other materials and tools would have made the Challenge easier to complete?
What human (group dynamics) problem were noticed?
If done again, what would you do differently?

Extensions:

Catapult design can be modified into a rocket launcher or a glider launcher.
Change size and shape of elastics to experiment/improve efficiency.
Challenge students to develop a catapult that will launch the marshmallow the highest; the farthest.
Draw a scale orthographic drawing of the catapult.
Vary the weight slightly, use something safe to catapult, other than a marshmallow.

Some Useful Resources:

Macaulay, David, *The Way Things Work.*
Photographs of successful catapults
The Metropolitan Toronto School Board, *By Design: Technology Exploration & Integration.*
Any high school physics textbook
The Metropolitan Toronto School Board, *Springboards to Technology.*
Corney, Bob & Dale, Norman, *Technology I.D.E.A.S.*

Integration Connections

Mathematics	Science	Technology
• measurement • linear, conversion of units • right angled triangles and Pythagorean theorem • rectangles and their properties • graph (release angle, distance traveled) — is there a relation here? • measure different angles of release to determine best angle	• elastic tension — force in N • elasticity — potential and kinetic energy • energy conversion/law of conservation of energy • investigate Newton's 3 laws of Physics, especially "every action has an equal & opposite reaction" • making predictions • materials investigation — the composition and properties of elastic materials; natural and synthetic • testing and changing variables, i.e. angle of launch; length or thickness of elastic; mass of object • can calculate work done by marshmallow after measuring force of elastic & using formula W-Fxd	• research it — library, text, CD ROM, Internet • safe & efficient use of tools • materials study, properties of • problem solving processes — use of chart • sketching, idea • prototype explorations • experimentation with materials • preparation and processes of materials to length and shape • assembly of materials in conjunction with plans using different joining & fastening of materials • efficiency — did it meet the objectives? • accuracy — what construction modifications are/were necessary to meet objective? • construction — the best use of the materials and tools • group dynamics — what problem were encountered in group assignments and working together? • include a written report

Teacher Talk:

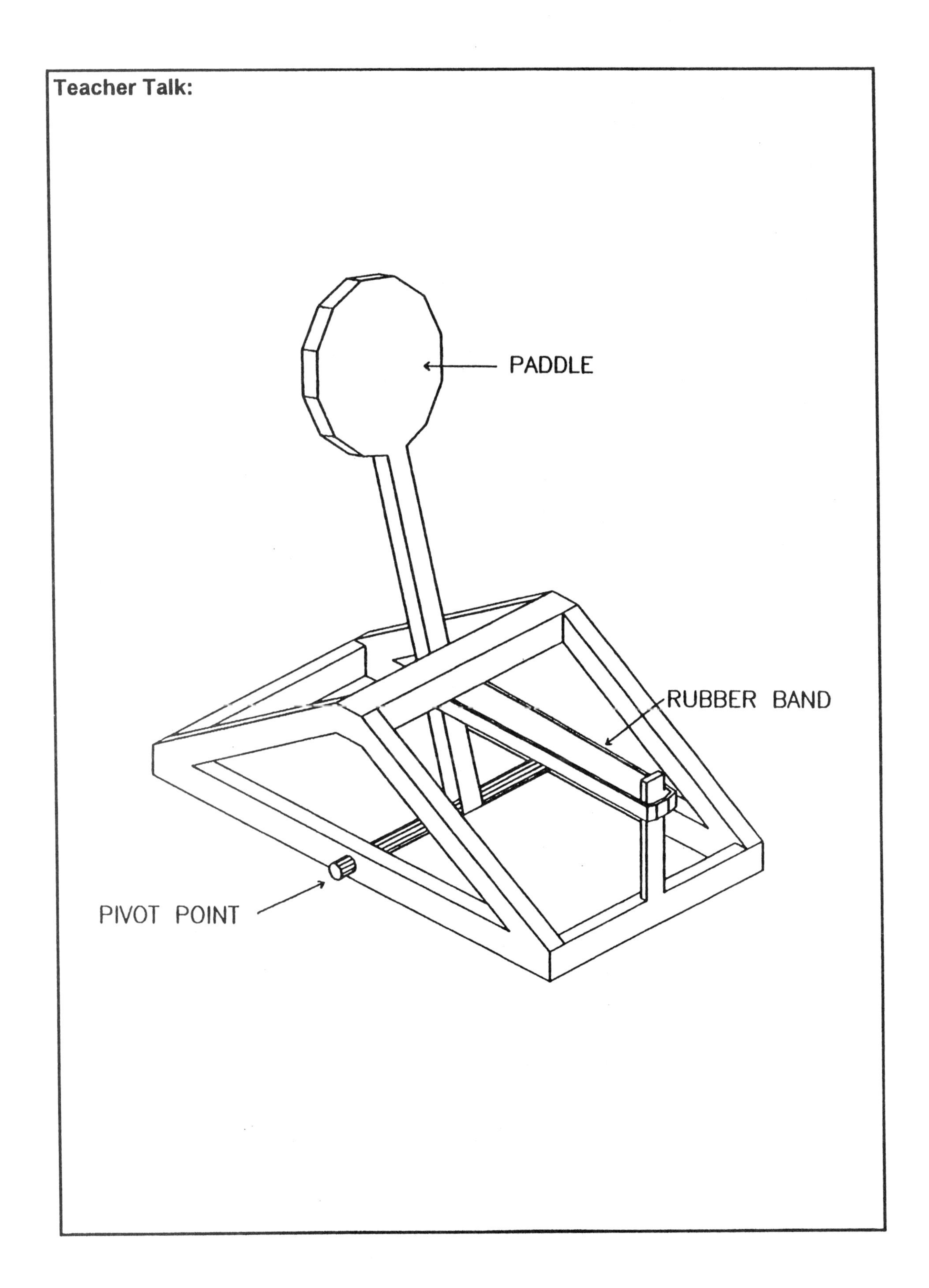

Theme:

FORCE

Title:

"Pass The Buck(et)"

Integration Connections:

See page 104.

Situation:
The early pioneer farm was a difficult place to work. Animals and children were used to doing much of the work on the farm including the grinding of grain to make flour.

Challenge:
Design and build a working model of a water wheel that will harness the motion of moving water or the gravitational force of falling water to turn the grinding stone in a mill.

Materials & Equipment:
Wood: basswood or willow strips 1 cm x 1 cm x 1 m and 0.5 x 1 cm x 1 m in length; 1/16"-1/4" dowel; Bamboo BBQ skewers; round wooden disks — 2-3" diameter
Tools: hack saw, miter box or miter saw, assorted drill bits, glue gun, 30 cm ruler, hand drill
Drafting materials: set squares, architects scale, T square, protractor, compass
Other: Lego 1030 Kits; plastic cups (35 mm film containers or substitute) worm gears
*Optional: hole saw, drill press, scroll saw, sanding drum if disks are to be manufactured

Parameters:
A maximum of 4 m of wood strip materials; maximum size of water wheel 30 cm x 30 cm.
Students are to work in groups of 3.
The device should be of durable construction.
The water wheel must rotate freely with minimum friction.
The water wheel must turn a gear to change vertical motion into horizontal motion.
Time frame 6 hours of class time.

Exploring Ideas:
Students will:
Research the design and purpose of historic water wheels;
Examine the positioning and operation of a water wheel in a river or water fall;
Examine the use of gears and different gear configurations to regulate speed and change motion direction;
Experiment and safely manipulate tools and materials in the most efficient manner.

Choosing & Building the Solution:

Students will:
Brainstorm ideas, produce rough sketches, negotiate the best solution, create a model or prototype, revise the model, select the final design, create working drawings, and construct the final machine.

Reflections:

Were all the safety considerations recognized?
Was the motion of the water harnessed by the water wheel and transferred into horizontal motion?
Was the construction technique successful?
What made a successful water wheel?
What other machines, materials, tools would have made the Challenge easier to complete?
What human (group dynamics) problems were encountered?
If done again, what would you have changed/do differently?

Extensions:

Examine rotation of a device using water power and apply principles to the design and construction of a wind machine.
Change size and shape of the water wheel (multi-sided).
Change size (diameter) of the gears; vary number of sprockets or teeth in the gears
- experiment with different gear constructions
- harness the created energy to do other "work" activities.

Draw a scale orthographic drawing of the water wheel.

Some Useful Resources:

Macaulay, David, *The Way Things Work*, Houghton Mifflin Co., Boston, Mass.
Lego Technic Kits (see Sample Supplies)
Any high school physics textbook

Integration Connections

Mathematics	Science	Technology
• find the center of a circle • measure angles • sum of angles in an n-gon = 180 (n-2) • construct a circle with a given radius & determine where the vertices for an N-gon should go • gear ratios • Pythagorean theorem (building supports) • volume of the cylindrical cups • vary size of containers and compare to the energy generated (graph info) • how does the number of sides affect the energy? • unit rate = number of revolutions/min.(sec.) • vary radius & compare energy generated	• gravitational potential, energy- converted to kinetic energy • definitions & understanding of simple machines — wheel & axle, gears changing direction of force, making work easier • Newton's laws of Physics • observe RPM's of water wheel • measure force of water turning the wheel (attach spring scale to a downward bucket & measure the resistance (force required to stop the movement) • volume of buckets — measure — set up controlled experiment to vary this & observe affect on RPM's	• research rotational motion in the water wheels • safe use of tools and equipment • selection of appropriate materials and combinations of materials • problem solving process • brainstorming • sketching • prototype making • preparation, shaping, and processing materials • assembly of materials — drilling, joining, fastening, assembly of gears and supporting structure • appropriate use of materials and the best construction* • efficiency-mechanical efficiency of the machine gears, low friction; rotating of the wheel *(strength of construction) • group dynamics — what problems were encountered in group assignment working together? • include a written report

Teacher Talk:

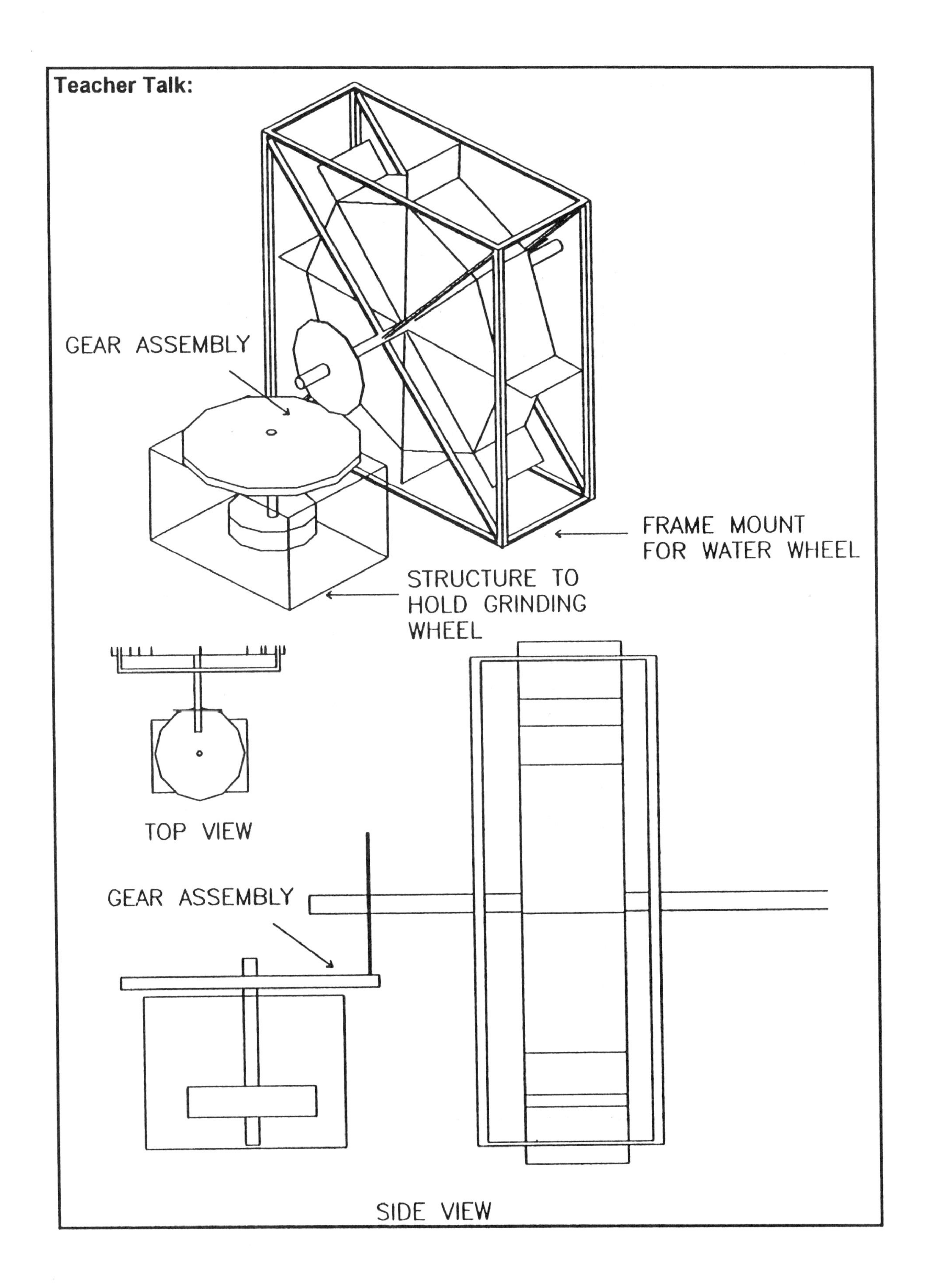

Theme:

FORCE

Title:

"Dare to Take the Leap"

Integration Connections:

See page 108.

Situation:

You are a member of a construction team. The owner of an amusement park has asked your team to design a structure to be used for a bungee jump over a lake.

Challenge:

Design and construct a stable structure which will support a diving board off which a range of weights attached to it can be dropped. The maximum recoil is desired without the bungee jumper getting wet.

Materials & Equipment:

Elastic/rubber bands, glue guns, or carpenter glue, bucket of water, various weights, Bristol board, miter saw, miter box, 2 cup hooks, counterweights, rulers, Newton spring scale, scissors, basswood, or willow strips 0.5 cm x 1 cm x 1 m and 1 cm x 1 cm x 1 m length

Parameters:

Must recoil at least 3 times.
Work in small collaborative groups of 3.
Height of structure 1.5 m (could be placed on table).
Minimum weight 300 g, maximum weight 450 g.
Use listed materials only.
Suggested building time: 5 hours in class.

Exploring Ideas:

Types of structures (tower)? Research designs of towers — Eiffel Tower, television, radio, electrical transmission towers.
Types of elastic bands, varying lengths of elastic bands and diving board?
Counter levers?
Tensile strength of elastic.
Tower construction: girders, lattice structures, guy ropes, bracket supports (beneath and above).

Choosing & Building the Solution:

Select a structure; decide maximum length of diving board which can be used; decide position of diving board; decide counter lever weight; decide type and length of bungee cord preferred.

Reflections:

Did the elastic perform as expected? If not, why not?
What problems occurred in the process? How were they handled?
Did the group members share responsibility?
If you were to do it again, what changes would you make?

Extensions:

Build higher towers.
Add more weights.
Investigate bungee jumps being used
Investigate alternative structures (popsicle sticks).

Some Useful Resources:

Techniques for Technology, 1989. (see Sample Suppliers)
Many universities have outreach programs. Call your local university or use the Internet to contact them.
Any high school physics textbook

Integration Connections

Mathematics	Science	Technology
• linear measurement • ratio of weight: stretch length • table of values for different lengths, weights • graph (weight, length) • type of relationship in a linear, slope equation • data analysis • extrapolation/ prediction • cost analysis • graph recoil lengths	• measuring weights with Newton Spring scale/mass with triple beam or equal arm balance • vary length or thickness of elastic within controlled experiments (Scientific Method) • hypotheses testing • observation • mass vs. weight • compare elastic length when weight is static vs. length when weight is dropped • understanding of gravity, tension, recoil, elasticity, stresses, cantilever • use Newton Spring scale to measure force (gravitational pull) to determine counterweight needed • center of balance, base of support Newton's Laws of physics, especially "every action has an equal and opposite reaction"	• research the design and construction of different tower structures • vary length of diving board (is there a max. length)? • material selection • examine the sport/activity of bungee jumping • examine elastics and weight holding capacity • sketching designs • experiment and safely manipulate tools and materials to the most efficient manner • preparation and processing of materials • assembly of materials - joining, fastening, gluing • examine structure stability • examine strength of the construction • can the tower accept the force of the falling object and retain its balance? • examine group dynamics — what problems were encountered? • include a written report

Teacher Talk:

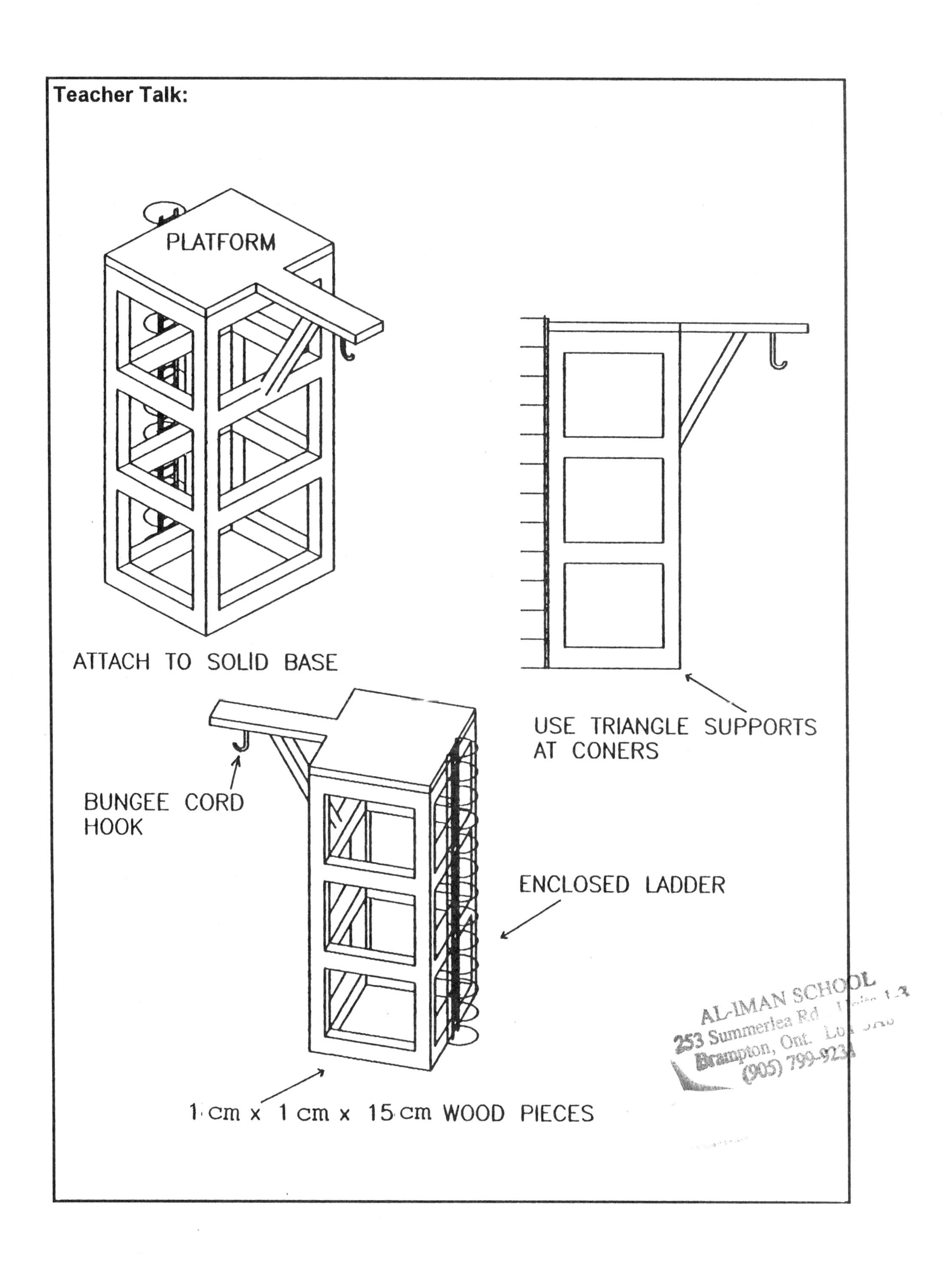

Environment Challenges

Introducing Challenges

Before you begin this section, please consider *Questions to Ask at the Planning Stage* on pages 21-23. Space is left below for your notes on how you might change your strategy next time or perhaps to record in which classes you have used each Challenge, etc.

Teacher Notes

Theme:

ENVIRONMENT

Title:

"Pet Palaces"

Integration Connections:

See page 114.

Situation:

You are an assistant at a local animal shelter and are upset with the living conditions of the animals in your care. Your boss asks you to help create better living conditions for the animals in your care.

Challenge:

Design and construct a more suitable habitat for the animal in your care.

Materials & Equipment:

masking tape	scissors/glue
glue gun	papier maché
coping saw	1 cm x 1 cm squared wood
hammer	Bristol board
wire mesh	applicator sticks
tubing	popsicle sticks

Parameters:

Students will work in small collaborative groups.
Habitat to accommodate one animal.
Habitat must encompass or satisfy animals' needs — food, movement, shelter, waste management, safety, water, breathing, sleep, hygiene, etc.
Availability and cost of materials must be considered.

Exploring Ideas:

Will the habitat meet all the animal's needs?
Will it be easy to maintain?
Where will your model or project be displayed?
How is your task to be divided up (division of labor)?

Choosing & Building the Solution:

Group will research and gather information to help determine the most suitable habitat for their animal.
Design and build the habitat.

Reflections:

Did it meet all the animal's needs?
What did you learn about your animal?
What would you do differently the next time to improve your solution?
What did you enjoy most about doing this entire project?

Extensions:

Create a classroom animal shelter.
Learn about animal care — field trip to zoo, pet store, local vet, animal shelter.
Career exploration — Biologist, Zoologist, Animal Husbandry.
Is your solution marketable?

Some Useful Resources:

Grant, Lesley, *Great Careers for People Concerned About the Environment.*
Czerneda, Julie, *Great Careers for People Interested in Living Things.*
Mason, Helen, *Great Careers for People Who Like Being Outdoors.*
Czerneda, Julie, *Great Careers for People Who Like to Work with Their Hands.*
The Metropolitan Toronto School Board, *Springboards to Technology.*
The Metropolitan Toronto School Board, *By Design: Technology Exploration & Integration.*
CD Rom — animals
School library
Visitations — zoo, animal shelter
Humane society
Local veterinary clinic

Integration Connections

Mathematics	Science	Technology
• linear measure • volume/mass • symmetry/asymmetry • utilization of space • scale drawing • mapping, grid work and planning • graph results	• researching specific animal needs • characteristics of living things • observation/inquiry/ questioning • recording data • making inferences and predictions • exploration of various materials, e.g., wood shavings, sawdust, astro turf, etc. • variables of habitat — food, bedding, odor control, absorption of excretions • field testing in the classroom	• designing — rough and final draft — scale drawing • choosing materials • open ended problem solving process • safe environment? • selection and safe use of tools • develop a list of materials • stages of construction • did finished product meet the plan requirements? • ergonomics — did the finished product help to improve the life and general well being of the animal being provided for?

Teacher Talk:

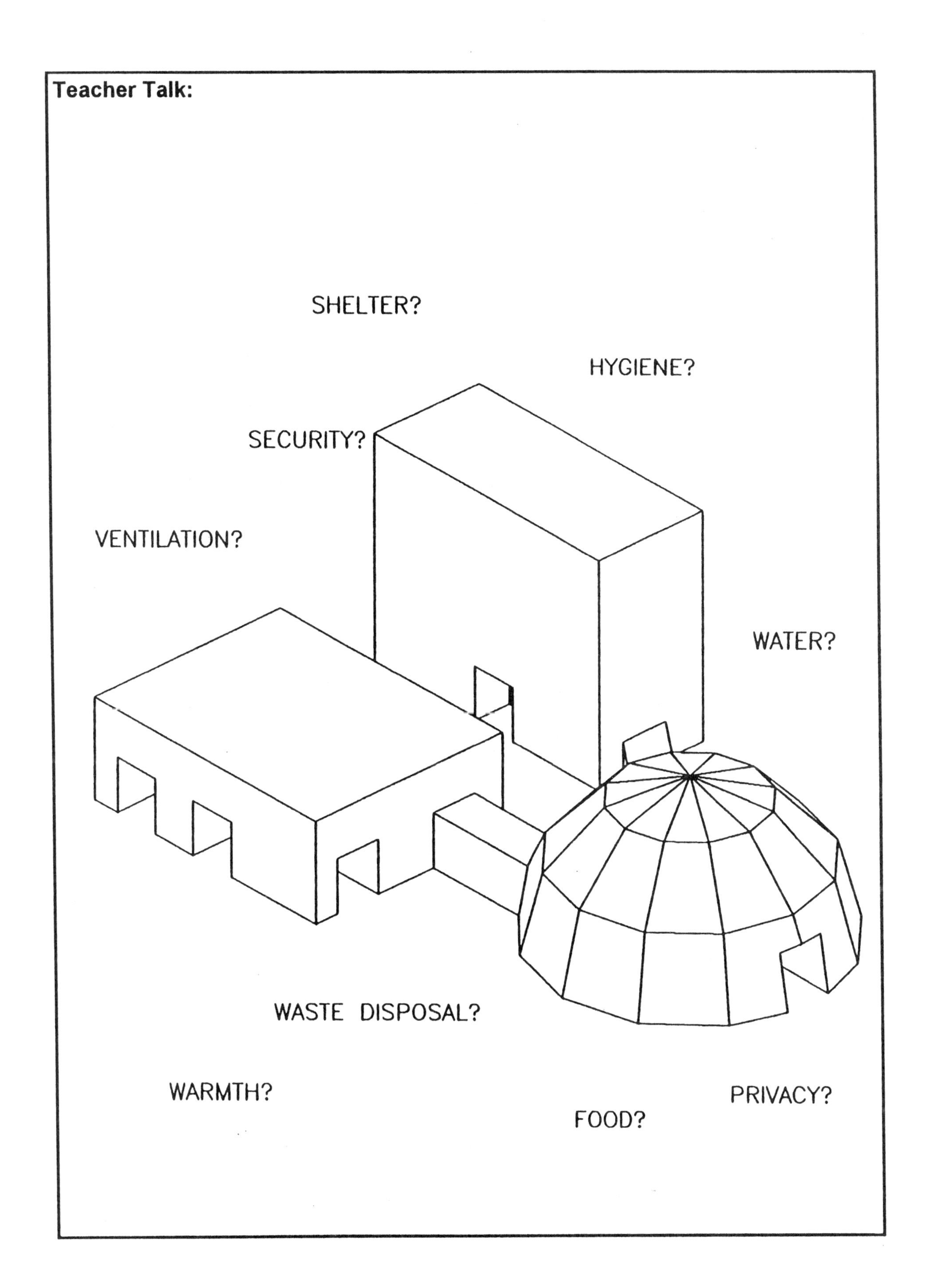

Theme:

ENVIRONMENT

Title:

"Live a Sheltered Life" (Survival Shelters)

Integration Connections:

See page 118.

Situation:

You're intending to go on a camping trip to a national park and must prepare for yourself a temporary shelter.

Challenge:

Design and create a shelter that will enable you and your party to survive. The month is January, and the park is in an area with a cold climate at this time.

Materials & Equipment:

any available and appropriate organic material
recyclable material — newspaper
glue
hammer
Bristol board
sewing machine
fabric
1 cm x 1 cm squared wood
scrap wood
coat hangers

Parameters:

Students will work in small collaborative groups.
Parameters for shelter — must be portable.
Availability of materials.
Scale of model — full size or model?
Shelter for 2, 3, or 4 people?
Students decide on how long the party will be in the shelter (e.g., 2 hours, 24 hours, or a full week).

Exploring Ideas:

Will this shelter meet your group's needs (e.g., size, warmth, comfort, moisture resistance)?
Insulation values?
Portability?
Moisture barrier — permanence?
Economy of space (shapes of structure).

Choosing & Building the Solution:

Research and examine the variables which will affect this shelter.
Design and build the shelter or a model of this shelter to scale.

Reflections:

Does the shelter meet the needs as outlined?
What did you learn about shelters?
Could you improve on your design? How?

Extensions:

Plan a camping trip that includes all the personal and safety needs for successful winter camping.
Visit a tent manufacturer.
Research foods in the natural environment (extension to survival in general).
Display in the gym.
Setup criteria for judging.
Write up an instruction manual on how to put the structure together.

Some Useful Resources:

A conservation area in your area
Library (books on shelters & structures)
Wilderness survival information from nature and hiking organizations
Novel study — Farley, Mowat, *Two Against the North*; George, Jean Craighead, *My Side of the Mountain*
Mason, Helen, *Great Careers for People Who Like Being Outdoors*.

Integration Connections

Mathematics	Science	Technology
• shape / 3D • volume • area • perimeter • surface area • scale drawing • geometric nets • build 3D shapes • graphing/recording data • rating scale	• research human body needs (life functions) • research variables - impermeability - insulation values - moisture barrier • analyze research • make predictions • field testing • criteria met?	• designing (rough and final drafts) • create a mock up • choosing materials • selection and safe use of tools • parts and materials list • stages of construction • design process • did the finished product meet your planned requirements?

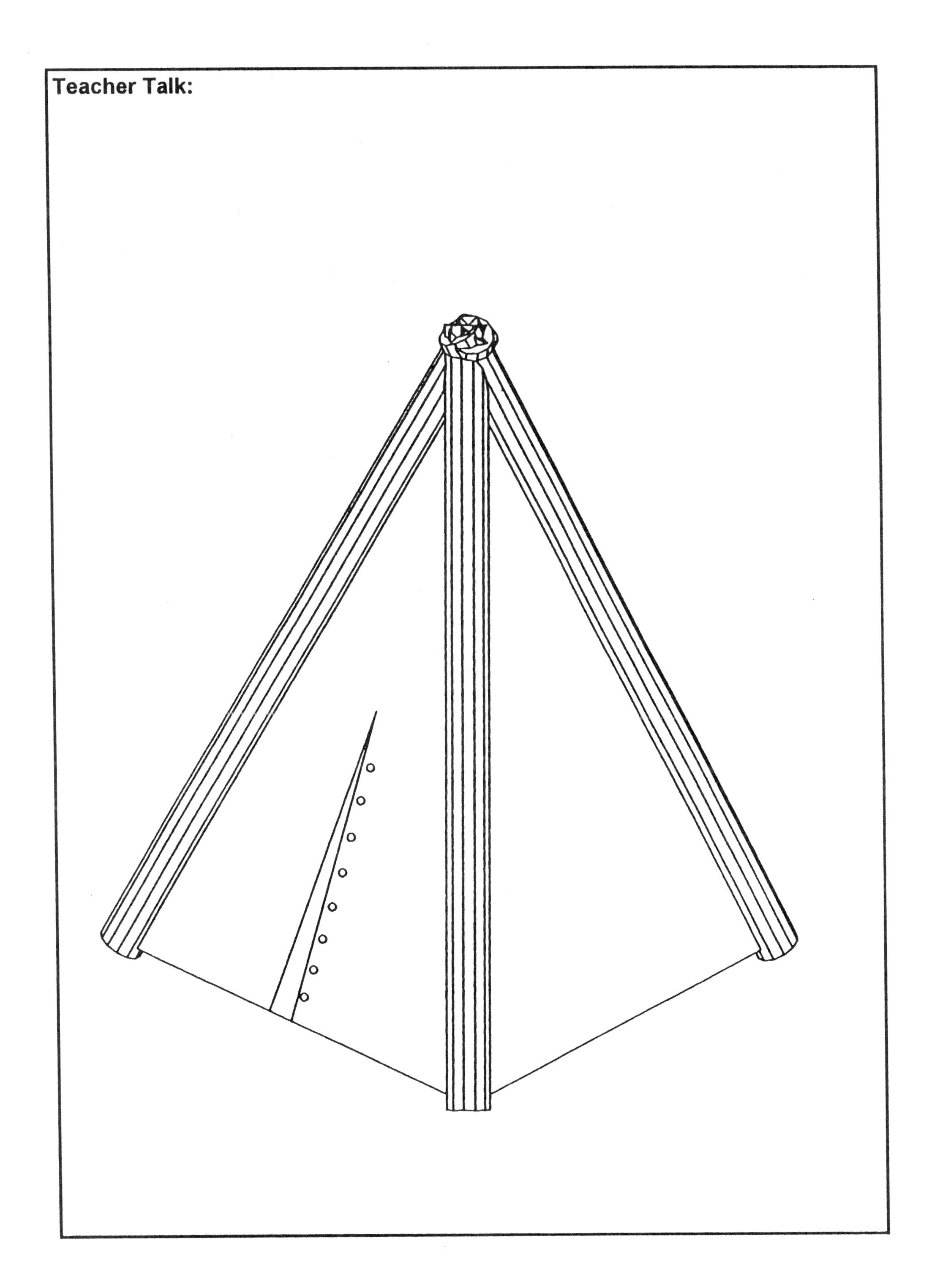
Teacher Talk:

Theme:

ENVIRONMENT

Title:

"Landscape for a Retirement Villa"

Integration Connections:

See page 122.

Situation:

You have been hired by the local retirement villa to redesign the grounds surrounding the villa itself. You want to improve the villa for the people living there.

Challenge:

You are a landscape architect. Design a landscape surrounding the villa that will be both accessible and inviting for all the people involved.

Materials & Equipment:

drafting materials
papier maché
paper
sand
plasticene

Parameters:

Students will work in small collaborative groups.
Model must be 1 m x 1 m.
Must encompass needs for the elderly.
Must be environmentally friendly.
Takes into account the seasons.

Exploring Ideas:

Guest speaker on landscape architecture.
Visit botanical gardens and local greenhouses.
Research horticultural concepts.

Choosing & Building the Solution:

The group will draw a grid map of the local villa landscape and redesign on a grid.
Submit a proposal to your landscape firm.
Presentation of proposal.
Build a scale model.

Reflections:

Would elderly people appreciate the landscape?
Is it accessible and safe?
Does it take into account the seasons?
Does it give people a variety of areas for privacy and social gatherings?
Is it safe for the disabled?

Extensions:

Students to write a plan for maintaining the grounds.
Write a maintenance schedule.
Map the present landscape of the local villa.
Each group to present their model to the class.

Some Useful Resources:

Czerneda, Julie, *Great Careers for People Interested in Living Things.*
Bartlett, Gillian, *Great Careers for People Interested in Art & Design.*
Mason, Helen, *Great Careers for People Who Like Working with People.*
Local retirement villa
Speakers that work with the elderly
Local horticultural center
Landscape architect
"Landscape 2.0" Microsoft
"Kid Cad" - Davidson, 3/D Building Kit: available at CompuCentre retail stores.

Integration Connections

Mathematics	Science	Technology
• area • formulas • grid design • curves • compass skills • protractors • squares • making a design with shapes • utilizing grid maps • scale map design • evaluate which shapes are most practical and cost efficient	• research perennials, annuals, shrubs, fertilizers • develop an organizer or graph that represents the differing flowering times of the plants and shrubs in your landscape • evaluate the organizer • does it communicate the information? • evaluate interviews and surveys	• explore drafting equipment • make plans • computer (CAD) • develop a list of materials • process planning • buying materials • processing and assembly of materials • how much does the landscape cost to maintain? • evaluate whether the landscape is practical and safe

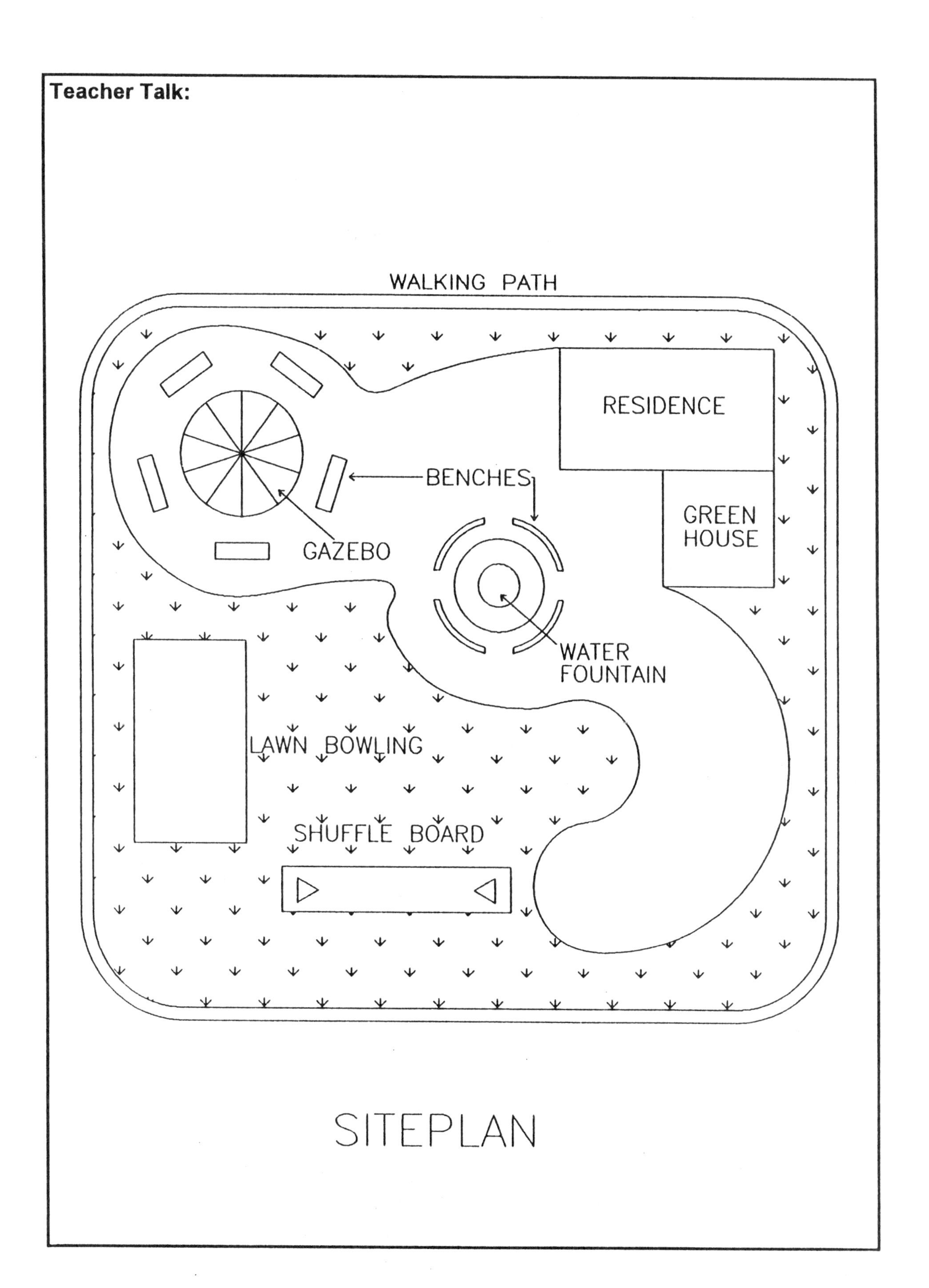
Teacher Talk:
WALKING PATH
RESIDENCE
BENCHES
GREEN HOUSE
GAZEBO
WATER FOUNTAIN
LAWN BOWLING
SHUFFLE BOARD
SITEPLAN

Theme:

ENVIRONMENT

Title:

"Wheels of Power/Water Works" (Alternate Energy Resources)

Integration Connections:

See page 126.

Situation:

A small town is troubled with a severe air pollution problem. The town is still growing and needs to use more energy in the manufacturing sector, without creating more air pollution. Water is your only constant and available source of power.

Challenge:

Design and build a simple machine(s) that uses water power to do work or perform a task for you.

Materials & Equipment:

simple machines (gears, pulleys and levers)

cups	paper cups
plastic cups	cord
coat hangers	elastics/rubber bands
source of running water	Lego
Tinkertoys	"Plastino" (eaves trough)
Meccano	buckets

Parameters:

source of water — sink, buckets, etc.
wheel and accessories to be portable
availability of materials

Exploring Ideas:

Research:
Alternate energy resources;
Pioneer technology (sluiceway, raceway);
Simple Machines;
Hydraulics;
Historical uses of water wheels.

Choosing & Building the Solution:

Designate the specific task your machine will perform.
Design and build it.

Reflections:

Did it perform the task?
What did you learn about power?
How could your design be improved?
What did you learn about simple machines?

Extensions:

Visit a local Pioneer Village.
Visit a hydroelectric power site in your area.
Go to a local science center to explore simple machines and water wheels.
Research tidal and wind power.
Mapping — waterways of your area.
Design a poster to advertise your invention.
Do a video commercial on your invention.
Research the effect of water flow on the natural environment (i.e., erosion, etc.).

Some Useful Resources:

Macaulay, David, *The Way Things Work.*
Grant, Lesley, *Great Careers for People Concerned About the Environment.*
Encourage research on inventors. For example, who was Rube Goldberg and what is he known for?
Library: CD ROM
Computer program: "Incredible Machines" (MS DOS)
Hydroelectric generating stations
Energy development corporations
Science textbook units on sources of energy
Rising, David, *Great Careers for People Interested in Film, Video, & Photography.*

Integration Connections

Mathematics	Science	Technology
• using formulas • rotation, circles • ratio • wheels • analyze results • discuss speed and volume of flow of water • graph results of work completed • graph results of tests • graph class results	• force, work & energy • water — water cycle • simple machines • levers and gears • hypothesizing • analyzing and graphing results	• test for suitability of materials • rotation building wheels • gears/gear ratio • simple machines • R.P.M. • water flow • build model and revise • are materials desirable for task? • does it work?

Teacher Talk:

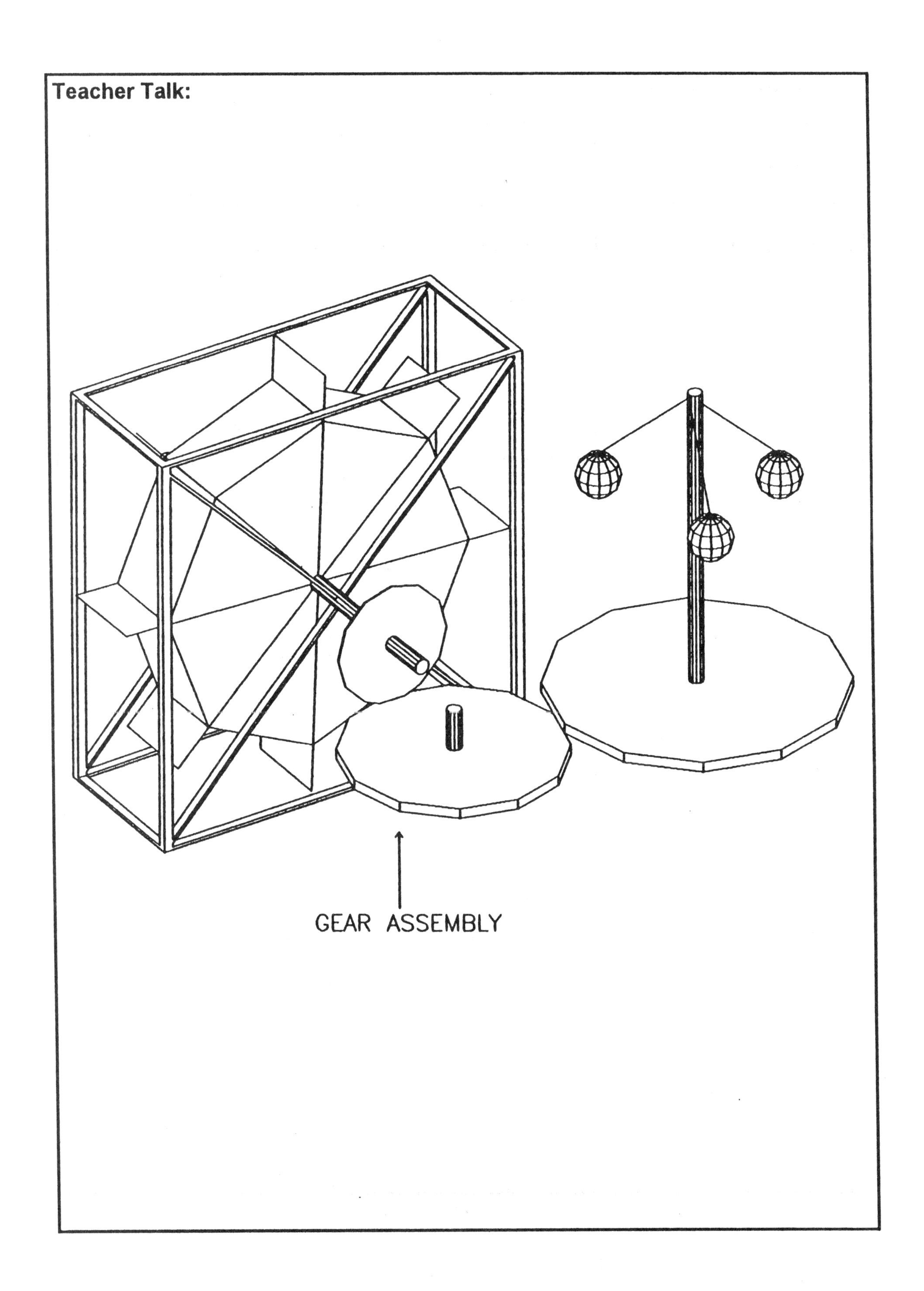

Theme: ENVIRONMENT

Title:

"A Place to Live" (Community of the Future)

Integration Connections:

See page 130.

Situation:

Your present community is in a state of transition. It is growing rapidly but does not appear to have a definitive plan for future development. As a member of the town planning department you have been asked to assess the present and future needs of your community and create a vision of a future community where the needs of all people will be met.

Challenge:

Design and construct a community as a satellite to your present community.

Materials & Equipment:

cardboard
Bristol board
model train ornaments
plexiglass/lexan
metal scraps
water
gravel
astro turf
glue
scrap wood
polyethylene
1 cm x 1 cm squared wood
scissors
earth
clay
simple tools as needed

Parameters:

Large group activity.
Organize into small groups.
Each group to consider:
- Availability of materials?
- Access to municipal offices?
- Cost of materials.
- Time.
- Needs of the community.
- Appropriate display area.

Exploring Ideas:

Look at "local community" researching as well into its historical roots.
Examine past and present needs.
Hold a public forum where all members of your present community give input as to its future needs.
Set up a planning department where you delegate responsibility as to the planning your satellite community — committees responsible for energy production and use, public works, natural resources, parks and recreation, education, medical facilities, water, etc.
Field trip to municipal planning department.

Choosing & Building the Solution:

Phase 1 Information gathering and research in small groups.
Each group to present its findings.
Phase 2 Reassign roles as to mapping/gridwork/plans; construction; write up/ report; presentation.

Reflections:

Does your community meet the needs of all age groups?
Do the structures in your community blend in with the environment?
Is your community "environmentally friendly"? How could you make it more so?
What did you find difficult about this project?
What would you do differently the next time?
How much do you think you contributed to this project?
Are all human needs met in the community you and your group developed?
What interests and skills do you have that would make you a good town planner? What additional skills would you need? What other careers are needed to keep a town going? Which ones of these would you be most interested in?

Extensions:

Group could present their class model at a business seminar or municipal presentation.
Group could write a business pamphlet to advertise their city.
Group could create a guided walking tour for their community, critique their community/ develop a rating scale.
Rate the collaborative skills of all classmates.
Group could present their model to town council.

Some Useful Resources:

Males, Anne Marie, *Great Careers for People Fascinated by Government & the Law.*
Mason, Helen, *Great Careers for People Who Like Working with People.*
Lang, Jim, *Great Careers for People Who Want to Be Entrepreneurs.*
V31633 "A Challenge — Plan a City of the Future"
Local archives and museum
CD-Rom: "Sim City" available at all stores selling computer simulations (see Sample Suppliers)
Municipal offices
Local and school library
Architects and town officials

Integration Connections

Mathematics	Science	Technology
• mapping grid/axis • scale drawing • distance (linear measure) • ratio and proportion • open-ended problem solving • geometry • accuracy of scale drawings	• research the environmental impact of cities and towns • how is power/energy transported? • gathering of information • decide on criteria for judging your model	• research tools and methods • plan the layout (plan alternatives) • choose the type of materials needed • plan view • scale drawing • perspective drawing/rendering • develop a materials list • determine a sequence for construction • choose a suitable location for the project • use a problem solving model such as - SPICE - PRAISE • review, revise, and modify

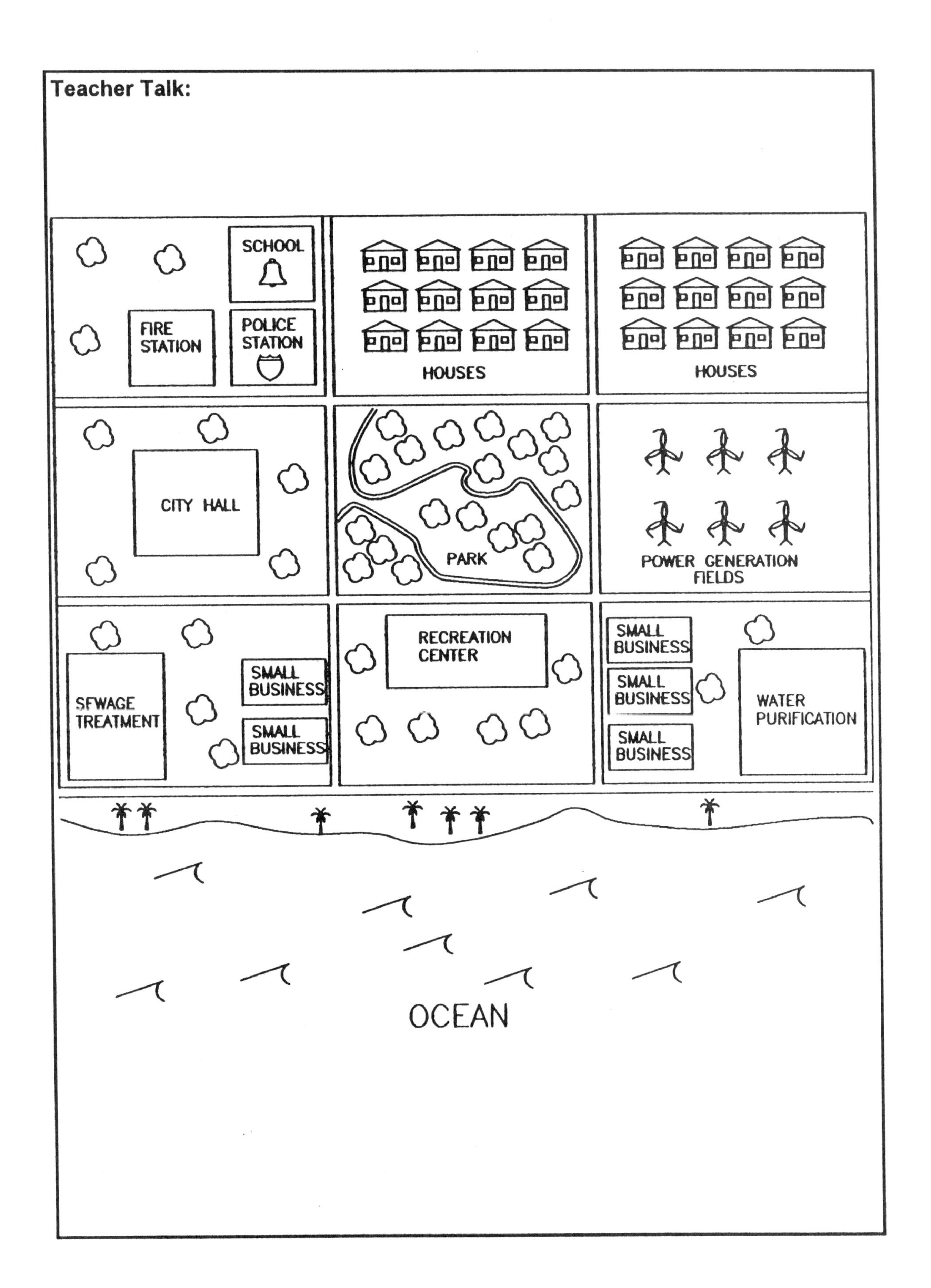
Teacher Talk:
SCHOOL
FIRE STATION
POLICE STATION
HOUSES
HOUSES
CITY HALL
PARK
POWER GENERATION FIELDS
RECREATION CENTER
SMALL BUSINESS
SMALL BUSINESS
SFWAGE TREATMENT
SMALL BUSINESS
SMALL BUSINESS
SMALL BUSINESS
WATER PURIFICATION
OCEAN

Theme:

ENVIRONMENT

Title:

Playgrounds with a Difference

Integration Connections:

See page 134.

Situation:

You have been asked by your school to create a playground for the very young children in a local school.

Challenge:

Design and create a model of a playground suitable for 3 to 6 year olds.

Materials & Equipment:

toothpicks
applicator sticks
1 cm x 1 cm squared wood
leather lacing
paper towel tubes
toilet tubes
rubber wheels
sand
rubber tires
Lego 1030
Popsicle sticks
string

Parameters:

Students to work in groups of 2-4.
Playground must be usable in all four seasons.
Playground should supply some challenges but at the same time be safe.
To have at least five different activities.
Indoor or outdoor playground.

Exploring Ideas:

Research what heights are suitable and safe for such a playground.
What are the abilities of the children for which this playground is being designed?
Find out what the preferences of these children are through detailed observation, and research.
Determine through research what materials will be most suitable for this project.
Visit other creative playgrounds and observe other design solutions.

Choosing & Building the Solution:

Decide on a focus for the playground.
Produce many possible design ideas and through the design process arrive at a final idea.
Create a diagram or plan of your proposed idea.
Decide on the scale and proceed to build a model of your idea.

Reflections:

Will children like your playground?
Will it be safe?
What will be the cost?
If the playground is outside, will it stand up to cold winters, or will it be useful in warm places like Hawaii or Florida only? How about places with a lot of rain like Vancouver, B.C., Seattle, Washington, or San Jose, Costa Rica?
Will it be suitable for older children? Up to what age?

Extensions:

Evaluate playgrounds in your community.
Market your ideas and see if you can sell your idea to the local Parks and Recreation.
Student designers interview 3 to 6 year olds to determine effectiveness of design.

Some Useful Resources:

Trip to the local park
Visit a daycare center
Arrange for a visit to a manufacturer of playgrounds
Resources in child development

Integration Connections

Mathematics	Science	Technology
• geometry — area, perimeter, cylinders, surface area • scale drawing • cost analysis • rating scale • accurate use of formulas	• estimate, predict, hypothesize • properties of matter • safety • mechanisms; simple machines • communicate findings • examine materials — mass, properties, and safety	• research appropriate materials • appropriate use of drafting equipment • plan and pictorial views and how to render them appropriately • putting ideas in diagrammatic form as per the design process • explore safe building "connectors" • build a model of the playground to a predetermined scale • were diagrams neat and accurate? • was the plan practical? • how realistic is the model? (checklists)

Teacher Talk:

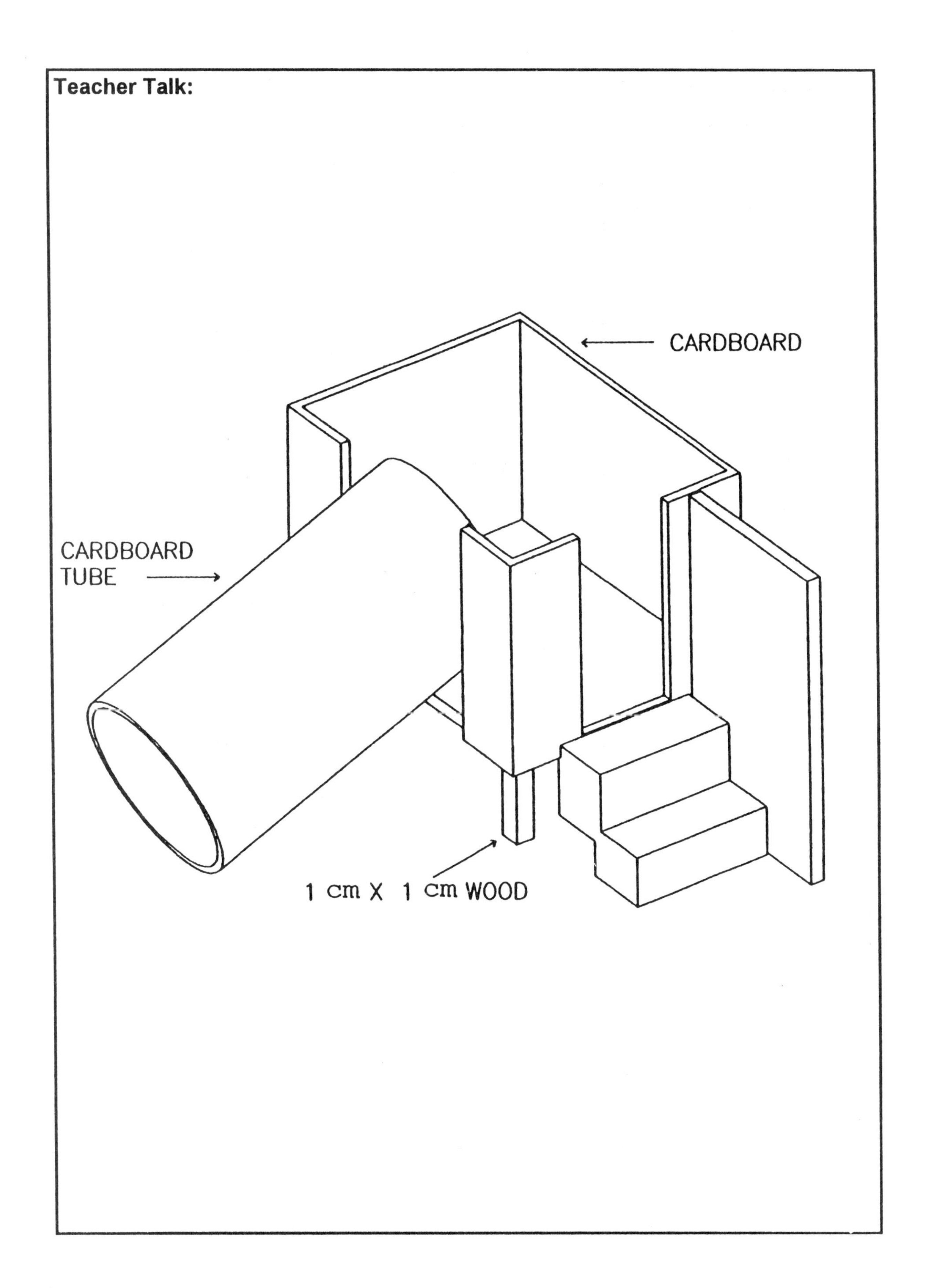

Useful Forms

Self-Checking my Work

Name: ______________________________ Date: ____________________

Circle the number that best describes your work on the Challenge you have just completed. Try to be fair and honest.

5	excellent
4	very good
3	good
2	satisfactory
1	poor

1. Effort

• I tried as hard as I could.	1	2	3	4	5
• I looked for help when I had difficulty.	1	2	3	4	5
• I helped someone else in the group.	1	2	3	4	5

2. Use of Resources

• I used books.	1	2	3	4	5
• I contacted people.	1	2	3	4	5
• I used technology.	1	2	3	4	5

3. Amount of Work

• I contributed to the group.	1	2	3	4	5
• I listened to others.	1	2	3	4	5
• I took responsibility for a task.	1	2	3	4	5

4. Use of Time

• I did not waste my time.	1	2	3	4	5
• I did not waste anyone else's time.	1	2	3	4	5
• I finished in the time allowed.	1	2	3	4	5

5. Other

• ____________________	1	2	3	4	5
• ____________________	1	2	3	4	5

Choose one item from the ones listed above in which you think you could improve. What could you do differently next time?

____

____

____

____

Evaluation Sheet for Small-Group Activities

Team: ______________________________ Date: ____________________

Agree on your answers as a group. Circle your choice.

1. Did we share?	YES	SOMETIMES	NO
2. Did we take turns?	YES	SOMETIMES	NO
3. Did everyone contribute?	YES		NO
4. Did we listen to each other?	YES	SOMETIMES	NO
5. Did we help each other?	YES	SOMETIMES	NO

Finish each sentence.

6. We agreed on ______________________________

7. We disagreed on ______________________________

8. We each had a task. The tasks were:

STUDENT NAME	TASK

9. We could improve by

____

____

____

____

____

Observing Others' Work

Name: ____________________ Date: ____________

Student who is being evaluated: ____________________

Circle the number that best describes the student's work on the activity just completed. Try to be fair and honest.

5	excellent
4	very good
3	good
2	satisfactory
1	poor

1. Effort

• Concentrated on the task.	1	2	3	4	5
• Looked for help when having difficulty.	1	2	3	4	5
• Helped someone else in the group.	1	2	3	4	5

2. Use of Resources

• Used books.	1	2	3	4	5
• Contacted people.	1	2	3	4	5
• Used technology.	1	2	3	4	5

3. Amount of Work

• Contributed to the group.	1	2	3	4	5
• Listened to others.	1	2	3	4	5
• Took responsibility for a task.	1	2	3	4	5

4. Use of Time

• Did not waste my time.	1	2	3	4	5
• Did not waste others' time.	1	2	3	4	5
• Finished in the time allowed.	1	2	3	4	5

5. Other

• ____________	1	2	3	4	5
• ____________	1	2	3	4	5

Choose one item from the ones listed above in which you think that this student could improve. Suggest what the student might do differently next time?

____

____

____

____

Some Useful Resources

Books

Career Connections Series, Trifolium Books Inc./Weigl Educational Publishers, Toronto, Ontario, and Calgary, Alberta, respectively.

An 18-book series, each 48 pp., softcover; career exploration integrated with activities in math, science, technology and other subject areas; interviews with 10 people in each volume.

Czerneda, Julie, *Great Careers for People Interested in Living Things*, 1993, ISBN 1-895579-00-7.

Edwards, Lois, *Great Careers for People Interested in the Human Body*, 1993, ISBN 1-895579-06-6.

Grant, Lesley, *Great Careers for People Concerned About the Environment*, 1993, ISBN 1-895579-04-X.

Mason, Helen, *Great Careers for People Who Like Being Outdoors*, 1993, ISBN 1-895579-10-4.

Richardson, Peter & Richardson, Bob, *Great Careers for People Interested in How Things Work*, 1993, ISBN 1-895579-08-2.

Richardson, Peter & Richardson, Bob, *Great Careers for People Interested in Math & Computers*, 1993, ISBN 1-895579-02-3.

Czerneda, Julie & Studd, David, *Teacher Resource Bank Series I*, 1993, ISBN 1-895579-40-6.

Filled with career and integrated curriculum activities throughout; tied closely to the above six volumes.

Bartlett, Gillian, *Great Careers for People Interested in the Performing Arts*, 1994, ISBN 1-895579-14-7.

Czerneda, Julie, *Great Careers for People Who Like to Work with Their Hands*, 1994, ISBN 1-895579-12-0.

Edwards, Lois, *Great Careers for People Interested in Sports & Fitness*, 1994, ISBN 1-895579-16-3.

Lang, Jim, *Great Careers for People Who Want to Be Entrepreneurs*, 1994, ISBN 1-895579-20-1.

Mason, Helen, *Great Careers for People Who Like Working with People*, 1994, ISBN 1-895579-18-X.

Rising, David, *Great Careers for People Interested in Film, Video, & Photography*, 1994, ISBN 1-895579-22-8.

Czerneda, Julie & Toffolo, Caroline, *Teacher Resource Bank Series II*, 1994, ISBN 1-895579-31-7.

Filled with career and integrated curriculum activities throughout; tied closely to the above six volumes.

Bartlett, Gillian, *Great Careers for People Who Like Art & Design*, 1996, ISBN 1-895579-48-1.

Czerneda, Julie & Vincent, Victoria, *Great Careers for People Interested in Communications Technology*, 1996, ISBN 1-895579-74-0.

Males, Anne Marie, *Great Careers for People Fascinated by Government & the Law*, 1996, ISBN 1-895579-50-3.

Mason, Helen, *Great Careers for People Interested in Food*, 1996, ISBN 1-895579-47-3.

Sommers, Jo Anne & Sharon, Donna, *Great Careers for People Interested in Travel & Tourism*, 1996, ISBN 1-895579-49-X.
Vincent, Victoria, *Great Careers for People Interested in the Past*, 1996, ISBN 1-895579-51-1.

Chapman, C., et al. *Collins Technology for Key Stage 3: Design and Technology, the Process.* London: Collins Educational, 1992.
Good ideas to assist teachers with the design process.

Corney, Bob & Dale, Norman, *Technology I.D.E.A.S.*, Maxwell MacMillan Canada, 1993, ISBN 0-02-954154-9.
Good ideas for projects for elementary students.

Glencoe, *Technology: Science & Math in Action Book 1*, ISBN 0-02-636945-1; *Technology: Science & Math in Action Book 2*, ISBN 0-02-636948-6, McGraw Hill, New York, 1995.
Good project ideas.

Hacker & Borden, *Living with Technology*, Nelson Canada, Scarborough, ISBN 0-8273-4907-6.
Good for Challenge "Welcome to the Fast Lane."

La Porte, *Technology Science and Mathematics; Connection Activities: A Teachers Resource Binder*, (Correlated to *Technology: Science & Math in Action Books*), McGraw Hill, 1996.

Macaulay, David, *The Way Things Work*, Houghton Mifflin Co., Boston, Mass. (also available on CD-ROM)
Useful as a student reference for several Challenges.

The Metropolitan Toronto School Board, *By Design: Technology Exploration & Integration*, Trifolium Books Inc., Toronto, Ontario, 1996, ISBN 1-895579-78-3.
Excellent ideas on technology education for middle schools; junior high schools.

The Metropolitan Toronto School Board, *Springboards to Technology*, Trifolium Books Inc., Toronto, Ontario, 1996, ISBN 1-895579-86-4.
A good teacher resource for elementary, middle, and junior high school.

Measuring Up: Prototypes for Mathematics Assessment. Mathematical Sciences Education Board, National Research Council. Washington, DC: National Academy Press, 1993; available in Canada only from Trifolium Books Inc.*
National Science Education Standards. Washington, DC: National Academy Press, ISBN 0-309-05326-9; available in Canada only from Trifolium Books Inc.*
Excellent ideas on science teaching and a good section on integration with mathematics and technology. National Academy Press; available in Canada only from Trifolium Books Inc.*

Heide, Ann & Stilborne, Linda, *The Teacher's Complete & Easy Guide to the Internet*, Trifolium Books Inc., Toronto, Ontario, 1996, ISBN 1-895579-85-6.

The book has wonderful features for teachers — "Getting Started" exercises encourage teachers to explore internet use for themselves, and "Project Ideas" provide many, many ideas for student use at all levels, from K-12. And at the end of the book — a bonus — an appendix of great educational resources that can be found on the Internet.

Williams, Peter & Jacobson, Saryl, *Take a TechnoWalk*, Trifolium Books Inc., Toronto, Ontario, in development, ISBN 1-895579-76-6.

Instead of a nature walk, take your students on a technowalk (or several of them). The book gives ideas on how to do it.

*For National Academy Press publications order from NAP in the United States. In Canada, send, fax, or e-mail your order to

Trifolium Books Inc.
238 Davenport Road, Suite 28
Toronto, ON, M5R 1J6
tel. (416) 960-6487
fax (416) 925-2360
e-mail: trising@io.org

National Academy Press
2101 Constitution Avenue, NW
Lockbox 285
Washington, DC 20055
Tel. 1-800-624-6242
Fax (202) 334-2451

Magazines

Childress, V.; LaPorte, J.E.; and Sanders, M.E. "The Technology, Science, Mathematics Integration Project: Technology, Science, and Mathematics Teachers Working Together." *TIES Magazine*, March 1994.

Dugger, W.E., Jr. "The Relationship between Technology, Science, Engineering, and Mathematics." *The Technology Teacher*, Vol. 53, No.7 (1994).

Hutchinson, P.; Davis, D.; Clarke, P.; and Jewett, P. "The Design Portfolio: Problem Solving." *TIES Magazine*, November/December 1989.

Johnson, S.D. "Research on Problem Solving: What Works. What Doesn't." *The Technology Teacher*, Vol. 8 (1994); 27-29, 36.

Ruiz, E.; Bolyard, G.; Travis, J.; and Pyles, R. "Demystifying MagLev." *TIES Magazine*, March/April 1991.

Videos and Videodisks

V31412 Eureka: Package 01, 530 EUR.
SERIES: Eureka - EPISODE 01. TVOntario, 1983. 25 MIN., Narrated, Color.
Contains the following programs: Inertia; Mass; Speed; Acceleration, Part 1 and Part 2.
GRADE LEVELS: Intermediate, Senior

V31413 Eureka: Package 02, 530 EUR.
SERIES: Eureka - EPISODE 02. TVOntario, 1983. 25 MIN., Narrated, Color.
Contains the following programs: Gravity; Weight vs. Mass; Work; Kinetic Energy; and Potential Energy. GRADE LEVELS: Intermediate, Senior

V31617 Eureka: Package 03, 530 EUR.
SERIES: Eureka - EPISODE 03. TVOntario, 1983. 25 MIN., Narrated, Color.
Contains the following five minute programs: The inclined Plane; The Lever; Mechanical Advantage and Friction; The Screw and the Wheel; and The Pulley. GRADE LEVELS: Intermediate, Senior

V33967 Eureka: Package 04, 530 EUR.
SERIES: Eureka - EPISODE 04. TVOntario, 1989. 30 MIN., Narrated, Color.
This package includes programs that illustrate the three states of matter, describes the motion of molecules and explains how temperature is measured. GRADE LEVELS: Intermediate, Senior

V31619 Eureka: Package 05, 530 EUR.
SERIES: Eureka - EPISODE 05. TVOntario, 1983. 45 MIN., Narrated, Color.
Contains the following five minute programs: Atoms; Electrons; Conduction; Volume and Density; Buoyancy; Convection; Heat vs. Energy; Radiation Waves; and the Radiation Spectrum.
GRADE LEVELS: Intermediate, Senior
Above series available from TVO in Canada and from MTV in the United States (see Sample Supplies, page 147)

The Puzzle of the Tacoma Narrows Bridge Collapse. A videodisc with 68-page instructional pamphlet; Product : TAC-01-PS: Available from Videodiscovery, Inc. 1-800-548-3472

Kelvin MagLev Racer™. Video #650323 (how to build and use MagLev tracks and cars). Kelvin Electronics, 10 Hub Drive, Melville, NY, 11747, fax (516) 756-1763; 1-800-645-9212

Useful Internet Addresses

Special Note about "The Web"

Netscape is presently the best Web software to enable you to contact a wealth of resources on the World Wide Web. Please observe your own school's security policies, before you assign students to get into the Internet. Once in Netscape, you will see a number of "search engines" additional pieces of software, embedded in Netscape, which will let you search for any and every topic you can imagine. Some of the search engines include "Web Crawler," and "Yahoo." Once underway, simply type in your topic, sit back and wait for the computers to talk to one another. This, in turn, will usually give rise to a number of resources and links to other sites and resources. Be forewarned: surfing the net can be addictive. Doing it on a school night might make it difficult to get to class the next morning![1]

Websites for a specific *Challenge*:

Websites for the *Challenge*, When the going gets tough, note the following:[2]

The following two websites contain accounts of projects by other students building Mars robots at other schools.

http://www.lunacorp.com/lcrover.htm
http://esther.la.asu.edu/asu_tes/TES_Editor/TESNEWS/4_VOL/No_2/re

Other websites of potential interest to students building *Challenges*

There is a Yahoo search page specifically of interest to K-12 students and teachers. It, like the main Yahoo search engine, allows users to search for information using the keywords supplied by the user, but this specific page allows the search to be confined to things that might be of interest to educators and students.[2]

http://beta.yahoo.com/Education/K_12

Educational Space Simulations Project. Includes: Space simulation "starter kit" for educators, student activities and experiments for use in space, launch and landing scripts.[3]

http://chico.rice.edu/armadillo/Simulations/simserver.html

Arizona Mars K-12 Education Program - brings Mars into the classroom with reading lists, missions, Mars facts and teacher's resources. Includes useful links such as: "Educational Resources at other WWW sites" "Planetary Science and Space Internet Links"[3]

http://esther.la.asu.edu/asu_tes/TES_Editor/educ_activities_info.html

Australia: A girls' science and technology high school has a webpage with links to some of their own experiments and projects in robotics and other things.[2]

http://www.ozemail.com.au/~mghslib/projects/mghsproj.html